Fundamentals
of **Intellectual Property Law**

Stephen McJohn
and **Lorie Graham**

Section of
Intellectual Property Law

AMERICAN BAR ASSOCIATION

Cover design by Best Design Chicago, Inc.

The materials contained herein represent the opinions of the authors and and/or the editors and should not be construed to be the action of either the American Bar Association or the Intellectual Property Section unless adopted pursuant to the bylaws of the Association.

Nothing contained in this book is to be considered as the rendering of legal advice, either generally or in connection with any specific issue or case. Nor do these materials purport to explain or interpret any specific bond or policy, or any provisions thereof, issued by any particular insurance company, or to render insurance or other professional advice. Readers are responsible for obtaining advice from their own lawyer or other professional. This book and any forms and agreements herein are intended for educational and informational purposes only.

19 18 17 16 15 5 4 3 2 1

Library of Congress Cataloging-in-Publication Data

McJohn, Stephen M., 1959- author.
    Fundamentals of intellectual property law / By Stephen McJohn and Lorie Graham.—First edition.
     pages cm
    Includes bibliographical references and index.
    ISBN 978-1-63425-253-9 (print : alk. paper)
    1. Intellectual property—United States. I. Graham, Lorie, 1962- author. II. Title.
    KF2980.M425 2015
    346.7304'8—dc23

                                    2015026289

Discounts are available for books ordered in bulk. Special consideration is given to state bars, CLE programs, and other bar-related organizations. Inquire at ABA Publishing, Book Publishing, American Bar Association, 321 N. Clark Street, Chicago, Illinois 60610.

www.ShopABA.org

To Ian

Contents

Preface

Are Neanderthal genes patentable? May a fan write *Harry Potter 8*? Could Amazon .com shut down Amazon Books, a women's bookstore? Is there copyright in a student's class paper? Why does so much copyrighted stuff stay up on YouTube? Can a video game company own its employees' ideas?

This friendly introduction to intellectual property gives the lay of the land. Not meant to be a book for lawyers or law students, it skips fine distinctions, like "primarily geographically nondeceptively misdescriptive trademarks." You can follow up, starting with the stuff in the "Sources" appendix.

About the Authors

Steve McJohn is a professor of law at Suffolk University Law School in Boston, Massachusetts, and is a specialist in intellectual property and commercial law. Professor McJohn received his B.A. in computer studies and his J.D., magna cum laude, from Northwestern University. Before law school, he was a software developer in the Radar Systems Group of Hughes Aircraft. After studying law in Germany and a federal appellate clerkship, Professor McJohn practiced law in the Chicago office of Latham and Watkins and taught at the IIT Chicago-Kent School of Law. He has authored numerous scholarly articles and textbooks on intellectual property, commercial law, computer law, artificial intelligence, and economic analysis.

Lorie M. Graham is a professor of law and co-director of the International Law Concentration at Suffolk University Law School in Boston, Massachusetts. She has been a visiting professor at Harvard Law School and at the University of Massachusetts-Amherst. She teaches courses on property law, international law, human rights, and indigenous peoples rights. Professor Graham holds a master of laws degree from Harvard Law School, and a B.S. and J.D. from Syracuse University. She has served as the executive director of the Harvard University Native American Program; practiced law at Kramer, Levin, Naftalis & Frankel in New York City; clerked for the Honorable Richard D. Simons of the New York Court of Appeals; and continues her practice in the area of human rights law. She recently published a second edition of *International Law: Examples and Explanations* (Wolters Kluwer, 2014), and has published widely on a range of issues, including articles on intellectual property, land claims, economic development, and media rights.

Copyright: What, Why, When, Whence?

*A day is coming, when, in the eye of the law,
literary property will be as sacred as whisky,
or any other of the necessaries of life.*[1]

Mark Twain

A tale of two copyrights

Charles Dickens lived in a world without copyright, part of the time. When Dickens was in the United Kingdom, copyright in his books protected him from copiers. But the United States, in those days, did not recognize copyright for foreign authors, so there Dickens could only watch as others freely published his books. Dickens lamented "the exquisite justice of never deriving sixpence from an enormous American sale of all my books."

To make money in the United States, Dickens had to go beyond book publishing. He did sell some books because some readers will always prefer the authorized edition, even if it costs more. Dickens also gave wildly popular public readings of his works, for a fee. He published his page-turner novels in chapters, which gave him a first-mover advantage. Eager readers waited at the port for each Dickens chapter rather than waiting longer for the cheaper, unauthorized version. His contemporaries Gilbert and Sullivan also struggled against copiers in the American market. To strike first, they staged premieres of *The Pirates of Penzance* close together in London and New York.

We now live in a world with more copyright.[2] Not only does the United States recognize copyright for foreign authors, but copyright is broader than in Dickens's times. Early copyright—for 28 years—applied only to books, maps, and charts. Now copyright applies to most creative works and may last for more than 100 years. We also live in a world with more copyright infringement. More things are copyrighted, and digital technology makes copying easy. The Internet is a global copy machine.[3] Accordingly, many of today's authors continue Dickens's strategies. Publishers rely on copyright, when it works, but, in addition, musicians and comedians sell tickets to performances, authors sell their work piecemeal, and movies open worldwide to get the jump on copiers.

What is copyright?

Theater student Ada writes a play, *Denial of Service*. She automatically owns the play's copyright. Others could infringe Ada's copyright by making *copies* of the play, *adapting* the play such as by making a movie, *distributing* copies to the public, *performing* the play publicly without her permission, or *displaying* copies to the public—although the display option is better for works like paintings. Copyright could be "Copyadaptdistributeperformdisplayright," and instead of ©, CADPD. The owner of the copyright in a book, a song, a movie, or a photo may sue infringers to get a court order to pay money and perhaps to cease the activity.

Copyright gives the legal right to prevent others from all of the following activities: copying, adapting, distributing, performing, and displaying.

Making copies of the work

Only the owner of the copyright in *Harry Potter and the Philosopher's Stone* can legally make copies of the book. Hence, *copy*right. Others could infringe the copyright by making a copy of a chapter or any part of the book; writing a book closely copying the plot and characters, even if no part is copied word for word; Xeroxing the book (copying in another form); scanning the book; and so forth.

Copyright has exploded in the digital age. First, electronic copies are easily made and easily passed along. Second, works in digital form, such as software, a digital photo, a book in PDF, or a song in MP3, must be copied in order to be used. To run software or play music or display a photo, a computer makes a temporary copy in its working memory. Consequently, simply

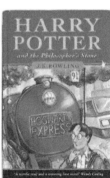

Potter Reproductions

using a work in digital form may infringe the copyright. This is a sea change: reading a physical book or looking at a painting does not involve making a copy, other than perhaps a mental copy.

Questions

1. How many ways could you make a copy of *Harry Potter*?
2. Booker copies four chapters from another author into a textbook, carefully giving full credit. Copyright infringement?
3. Wordwerf copies a poodle photo off a website for the cover of his book. The photo was freely available on the Web. Copyright infringement?

Answers

1. Well, you could copy it in longhand; Xerox it; scan it; take a picture of each page with a pinhole camera; save the text in PDF, ASCII, Braille, or Morse code; save a copy on a hard disk, flash drive, or cloud drive; quill it on parchment; or encode it in DNA. Consider some of the synonyms for *copy* found on thesaurus.com and the like: carbon copy, cast, clone, counterfeit, ditto, facsimile, forgery, hard copy, image, impression, imprint, likeness, microfiche, mimeograph, miniature, mirror, model, pattern, photocopy, photograph, photostat, print, replica, reprint, reproduction, rubbing, simulacrum, simulation, tracing, transcript.

   You could also copy the book less closely, such as by copying the characters and story, but not word for word.

   No doubt you can think of more ways to copy. We copy a lot, so the subject of copyright comes up a lot.

   What if you memorized the book? That's making a copy in a sense—but not in the sense of copyright law. Otherwise we would be infringing copyrights constantly. The law is not a hobgoblin for logical consistency.
2. Yes, this would be an example of infringement. Only the author has the right to make copies, whether or not proper attribution is given.
3. This is copying, and so it is likely infringement. The very point of copyright is to encourage authors to make their works public by giving them legal control over copying.

Adapting the work

The *Hitchhiker's Guide to the Galaxy*, *HHGG* for short, is a good example of a work that has been extensively adapted. A radio program begat *HHGG* the novel, and the novel begat a four-book trilogy, which begat *HHGG* the movie, along with *HHGG* audiobooks, translations, the *HHGG* game for the Commodore 64 computer, and, according to a Hungarian phrasebook, *Galaxis útikalauz stopposoknak*. The adaptations have outlived the author, Douglas Adams. His estate hired a kindred spirit, Eion Coifer, to write *And Another Thing*.

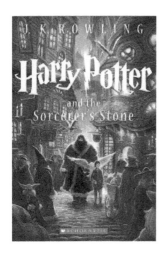   

Potter Transformations

The movie industry likewise thrives on adaptations. There's an Oscar for the Best Adapted Screenplay, usually from a book. Sequels are common. At the time of this writing, *Madagascar 3*, *Men in Black 3*, *The Hobbit*, an adaptation of a book and prequel of movies, and *The Avengers* (ditto) are out there. Broadway's biggest box office is in musicals adapted from movies: *The Lion King*, *Spamalot*, *The Producers*. Arthur Conan Doyle reluctantly resurrected Sherlock Holmes to keep a lucrative series going.

To use *Harry Potter* as our example again, others could infringe the copyright if, without permission, they made an Icelandic translation, *Harry Potter og viskusteinninn*, or an American translation, *Harry Potter and the Sorcerer's Stone* (changing "barking" to "off his rocker" and "Sellotape" to "Scotch tape"[4]), a Broadway musical, a movie, a ballet, an audiobook, a video game, or *The Annotated Harry Potter*.

The adaptation right gives the author some control over the artistic fate of her work. J. K. Rowling chose to stop after *Harry Potter and the Deathly Hallows*. One who wrote *Harry Potter Goes to CalTech* could nevertheless infringe. By the same token, the breadth of the adaptation right makes some wonder whether copyright goes too far.[5]

Having legal control over adaptations may affect how an author creates. J. K. Rowling, in writing the first *Harry Potter*, knew that only she would have the right to write sequels or have movies made. A subtle one: the "Bergman effect": when films are subtitled, only so many words per second may be shown. Swedish film director "Ingmar Bergman made two quite different kinds of films—jolly comedies with lots of words for Swedish consumption, and tight-lipped, moody dramas for the rest of the world."[6] The wordy comedies did not need subtitles for domestic audiences, while the laconic dramas could be subtitled for the world.

Questions

1. What adaptations were or could be adapted from the following? Feel free to speculate if you are not familiar with the work:
   The Hunger Games, the novel
   Windows 7, the computer operating system
   Madagascar, the movie
   Angry Birds, the video game
2. What's the oddest adaptation you can think of?
3. Shania Twain is thrilled to sell the copyright to her first novel, *A Is for Alphabet*, to Elastic Books. Encouraged, she starts the sequel, *B Is for Binary*. She also dreams of selling the movie rights to the books. Any problem?
4. What if copyright did not cover adaptations?

Answers

1. Following are the adaptations that were or could have been made:
   The Hunger Games was adapted into a movie, translations, and sequels. One could also adapt it into a play, an annotated edition, or a game (video or board).
   Windows 7 was adapted into *Windows 8*—and itself is one of a line of adaptations going back to *MS-DOS*.
   Madagascar was followed by sequels, and no doubt video games, books, and violin sonatas.
   Angry Birds was adapted into many more games, but also a movie, showing adaptation can flow any direction. It *could* likewise be adapted into any of these forms: sonnet, musical, or annotated scholar's edition.
2. Many possibilities exist here. *Pirates of the Caribbean*, a ride at Disneyland, was adapted into a movie, which then, of course, had sequels. Pierce Anthony's *Total Recall* novelized the movie *Total Recall*,[7] which was based on Philip K. Dick's story, *We Can Remember It for You Wholesale*.

   George Seurat's painting *A Sunday Afternoon on the Island of La Grande Jatte* (1884) was adapted into a musical, *Sunday in the Park with George*, by Sondheim and Levine.

A Sunday Afternoon on the Island of La Grande Jatte (1884) by George Seurat

Anything can be made into a musical. But not a patent, yet. Here's a possibility:

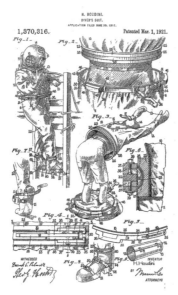

Having thus described my invention, I claim as new and desire to secure by Letters Patent:—

1. A diver's suit, comprising upper and lower body sections having air-tight and water-tight incasing means adapted to incase the diver against the escape of air from within the suit and against inflow of water when submerged, the sections having their

Diver's Suit, the Musical, Based on a Patent by Harry Houdini?

3. Problem: Elastic Books, now the owner of the copyright in *A Is for Alphabet*, has the right to adapt the book, such as making a sequel or a movie, or an Icelandic translation, *A er fyrir Stafróf*. Twain could infringe if she proceeds without Elastic's permission. Elastic may well agree, for a percentage. But Twain would have been wiser, had she but known, to sell only part of her rights.

Copyrights have been sold cheap. The authors of *Superman* sold the copyright for $130 in 1937. *Superman* and its many adaptations made hundreds of millions of dollars. Some 75 years later the author's estates were able to get the remaining rights back.

4. Would authors behave differently? J. K. Rowling would still have written *Harry Potter*, but perhaps differently if she had no right to make sequels and movies. She could have put it all in one big book, or wrapped everything up in the first. With no derivative right, she could not have prevented others from writing *Harry Potter 2* or making a *Harry Potter* movie—but audiences would likely prefer

J. K. Rowling's version, just as they prefer a genuine Monet even though the originals are no longer under copyright.

If copyright did not cover adaptations, it also would limit the author's artistic control. Art Spiegelman has staunchly declined to allow a movie to be made from his graphic novel *Maus*, saying: "I don't understand why everybody in this culture seems to believe it's not real until it's turned into a movie."[8] If he did not have the adaptation right, anyone could make a movie from *Maus*. Opinions vary on whether authors should have this control or whether others should have freedom to build on the work. What do you think?

Distributing copies of the work to the public

Harry Potter and the Deathly Hallows, Embargoed until Release Time (2007)

Nzgabriel, under http://creativecommons.org/licenses/by/2.5/legalcode

Under "First Sale" Doctrine, Amazon May Distribute Copies of *Harry Potter*

The distribution right makes it easier to show infringement. Were copyright limited to the right to make copies, a person simply selling unauthorized copies would not be infringing unless it could be shown that the seller made the copies.

A student selling used textbooks on eBay does not infringe the distribution right. The *first sale* doctrine limits the distribution and display rights to the owner of an authorized copy, so someone who owns a lawfully made copy may distribute it. A buyer of *Harry Potter* may sell his or her particular copy. By selling a copy, the copyright owner has lost control over the fate of that particular copy. First sale allows used bookstores to sell copyrighted books, libraries to lend copyrighted books, and museums to display or sell copyrighted paintings. But first sale applies only to the distribution and display rights. In other words, first sale allows a fan to sell her authorized copy of *Harry Potter* but not to make a copy, which would infringe the right to make copies, or to write a sequel—and that, in turn, could possibly infringe the adaptation right (we say "could," because, as we will see, it may be fair use if noncommercial).

The digital world raises two big questions about the scope of the first sale doctrine. First sale allows the owner of a copy to sell *that* copy. So one could deliver a DVD or USB storage device with the copy, but how could it be sold online? If I e-mail a copy to the buyer and destroy my copy, is that equivalent to selling my copy, which I may do? Or have I made another copy and distributed it, both actions potentially infringing?

Also, first sale rights apply only to the owner of a lawfully made copy. Software is often sold under a license agreement, which says the buyer gets a license to use the software but does not become the owner of a copy. So the buyer cannot legally sell her copy.

Questions

1. Which could infringe?
   a. Flash buys a copy of *Harry Potter*, scans it, and sells 50 copies around town.
   b. Madoff advertises cheap first editions of *Harry Potter* on eBay. He takes buyers' money and delivers nothing.
   c. Gordon buys 50 genuine copies of *Harry Potter* at an auction and sells them all.
   d. A library lends the same copy of *Harry Potter* 50 times in one year.
2. At an auction, Turing sells her computer, which is loaded with authorized copies of software. Has she infringed the distribution rights in the various copyrights, or does first sale apply?
3. Startup ReDigi allows people to resell songs from iTunes. ReDigi ensures that only one copy is delivered and that the seller's copy is destroyed. Would first sale permit this as the sale of a copy?[9]

Answers

1. Answers here are mixed:
   a. Flash could sell his authorized copy, but first sale does not allow him to make more copies or sell them.
   b. Madoff has not distributed anything, so he is not infringing copyright, but he may have to pay for fraud, wire fraud, false advertising, and plenty of other things.
   c. Gordon is not infringing the distribution right, because first sale allows him to sell the books.

d. First sale authorizes the owner of an authorized copy to sell,
      lend, or lease it. Without first sale, used bookstores and libraries
      would need permission from publishers.
2. Under first sale, Turing may sell authorized copies that she owns. But
   the software was probably provided under licenses that state that the
   copy does not belong to her and may not be resold. If the licenses are
   effective, Turing could be infringing. Do you think that's fair?
3. This looks like infringement: the making and distribution of more
   copies. First sale would permit only the sale of a particular copy.
   Could fair use apply? Stay tuned.[10]

Performing the work in public

Marketing Performance Material

Bar bands could infringe copyright by performing Beatles songs without permission, such as by paying for a license. The bar could likewise infringe if it
played recorded Beatles music over the sound system.

The copyright holder has the right to publicly *perform* the work, broadly
defined as "to recite, render, play, dance, or act it, either directly or by means
of any device."[11] Singing "25 Minutes to Go" privately at home will not
infringe Shel Silverstein's copyright, but singing the song at a concert could.

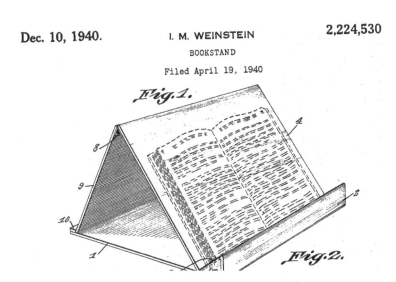

Dec. 10, 1940.   I. M. WEINSTEIN   2,224,530
                     BOOKSTAND
               Filed April 19, 1940

Handy Device for Performing a Literary Work

A public performance of a work may use a device. A coffee shop could infringe by playing a recording of "25 Minutes to Go" over the sound system. What about first sale, which limits the copyright owner's rights once a lawful copy is sold? First sale allows the buyer to distribute or display the copy but not to make copies, adapt the work, or perform it publicly.

How would Shel Silverstein enforce such a right? Shel's ghost could parade up and down every street in the country listening for unauthorized performances of his songs. Or someone could do it on his behalf. Collective rights organizations such as the American Society of Composers, Authors, and Publishers (ASCAP) or Broadcast Music, Inc. (BMI) represent considerable numbers of musicians. In fact, ASCAP and BMI do send agents to visit bars, coffee shops, restaurants, concert halls, and other venues. Crafty agents may phone businesses to hear if music is playing in the background or played to callers on hold. If music from their long list of copyrighted works is performed, the establishment will soon hear from the rights organization. That can be a surprise to a little coffee shop owner who thought she was entitled to play the music she purchased. Various licenses are available, however.

Questions

1. Which of the following could potentially infringe?
   a. Mick sings a Beatles song in the shower.
   b. Mick sings a Beatles song at a bar mitzvah.
   c. Mick sings Beatles songs onstage at the company picnic.
   d. A company pipes Beatles songs into the employees' cubicles.
2. From *The Boston Globe*: A resident of New Hampshire (state motto: Live Free or Die) was arrested 4 times in 26 hours for blasting the neighborhood with AC/DC's song "Highway to Hell" along with throwing a frying pan. She was charged with disturbing the peace and with assault. Was this copyright infringement as well?
3. Why don't waiters sing "Happy Birthday to You"?

Answers

1. Answers here are mixed:
   a. A song in the shower is the classic example of a nonpublic performance.
   b. As long as just family and friends are present, a song at a bar mitzvah is not a public performance.
   c. Even if Mick is not a professional, a company picnic would be a public performance (this looks like fair use, as do some other examples—stay tuned).
   d. Performing the music using a device is potentially infringing—and ASCAP and BMI are in the business of identifying such infringement and selling licenses. Businesses may either pay up or turn to less costly alternatives such as classical music, which is too old to be copyrighted, or MUZAK and so-called "easy listening" music.
2. A public performance of AC/DC's copyrighted music could possibly be copyright infringement. One need not sing the song to infringe. Playing a recording to the public is a public performance.

This was definitely public, hence disturbing the peace. But it may be protected by fair use (discussed later in the book) because she charged no fee and no lost licensing revenue is apparent. Anyway, AC/DC will be OK with it.

3. Restaurants prefer not to pay the licensing fee.

Displaying the work to the public

An artist built an installation in the Massachusetts Museum of Contemporary Art called *Training Ground for Democracy* in which visitors take on the roles of immigrants, activists, looters, and judges working their way through sets ranging from a movie theater to an aircraft fuselage. The artist and museum fell out, a common theme in copyright disputes. After the artist left for Europe, the museum proposed showing the artwork to the public but could not do it without infringing the artist's copyright, which includes the right to display the work to the public.[12]

The public display right matters little for many categories of works. No one cares too much about displaying the lyrics or musical arrangement of "25 Minutes to Go." Sound recordings cannot be displayed. Even the right to display such visual works as paintings and sculpture mattered little until recently. First sale applies to the display right, so only the owner of a painting, a poster, or a sculpture can legally display it. Displays of unauthorized copies of paintings, posters, and sculptures would infringe, but these displays have sparked few disputes. However, the Internet now permits displays of images everywhere, so public display right has become more important.

Question

An artist buys a book of photographs, cuts them out, hangs them in a gallery, and sells them. Infringement? Does it matter if the artist gives credit to the photographer? What if the artist draws on the photos?

Answer

The artist owns authorized copies of the photographs, so he may display them and sell them under the first sale doctrine. Failure to attribute makes no difference to copyright infringement.

Drawing on the pictures may infringe the adaptation right. But fair use may protect the artist.

Why does this book keep saying "could infringe"?

Someone who copies from *Harry Potter* or distributes copies, adapts the work, performs it, or displays it publicly could possibly be infringing copyright, but there is no infringement if she simply writes a different book about wizardly students—because copying ideas is not infringement. Nor is there infringement if a person makes fair use of the work, such as by writing a parody that comments on *Harry Potter*, or Xeroxing a passage to hand out to a copyright class, or writing a sequel for fun, without publishing it. For brevity's sake, this book says "could infringe" instead of repeatedly saying "infringes, unless fair use, the nonprotection of ideas, or some other limit on copyright applies."

Why copyright?

Laws forbid antisocial acts like murder, kidnapping, and double-parking. Copyright law forbids sociable acts like distributing books, singing songs, and translating *Harry Potter* into Icelandic. The purpose is to encourage people to distribute books, sing songs, and translate *Harry Potter* into Icelandic. How so?

Two reasons are most often given for copyright. The first is that copyright gives people an incentive to create works. The second is that copyright lets artists control use of their works and prevents others from distorting or knocking off those works.[13]

The first reason for copyright is economic. Absent copyright, a potential author might think, "Why take a year to write a book/make a movie/record an album if anyone else could then sell or give away copies of my work?" To give authors an *incentive* to create works, copyright law protects them against free-riders in that only the author has the right to exploit the work. The U.S. Constitution states such a reason for copyright and patent:

> The Congress shall have Power . . . to Promote the Progress of Science and the useful Arts, by securing for limited Times to Authors and Inventors the exclusive Right to their respective Writings and Discoveries.[14]

Even in industries like music, publishing, and software, many works would be created even in a world without copyright. But copyright increases the supply of works. Samuel Johnson observed, "No man but a blockhead ever wrote, except for money."[15]

Copyright also balances the right of authors to control their works against the interests of others in using the work. If copyright were merely an incentive, it would apply only to certain categories of works and would have real limits on duration. But copyright automatically attaches to any work, even those that would be created without copyright, such as diaries, academic works, free software, hobbyist works, student papers, and artworks—and lasts 70 years past the author's death.

The Virgin and Child with St. Anthony Abbot and a Donor (1472) by Hans Memling

Financial spurs to creativity go well beyond copyright.[16] Academics must, as the saying goes, "publish or perish." Students pay to be required to produce work. Advertisers produce quite creative works. Long before copyright law, the painting by Memling shown here was funded by the kneeling merchant so that he could be pictured in sacred company. A movie or opera may cast someone beloved of an investor. Patrons (like the donor in Memling's painting) have supported artists, as have governments and universities. Some musicians and comedians look more to sell tickets than recordings. Crowd-funding, such as Kickstarter, now allows authors to get broad-based financial support for proposed projects. The Nigerian movie industry relied little on copyright in becoming the second most productive in the world, between Bollywood and Hollywood.[17] Prizes, such as Nobel, Pulitzer, and Booker, plus publication, peer esteem, and self-fulfillment all fuel creativity—and apparently always have: scientists have found creative artifacts dating from pre-human hominids.[18]

Question

If copyright were just an incentive to produce works, how long would the term of a copyright be?

Answer

Consider this: a patent's term of about 17 years motivates plenty of inventors. An author's decision to create would rarely be affected by whether her copyright would expire in 20 or 30 or 100 years. How long do you think copyright should last?

When does copyright expire?

Boy Texting in 1911?

Neta Snook and Amelia Earhart (1921)

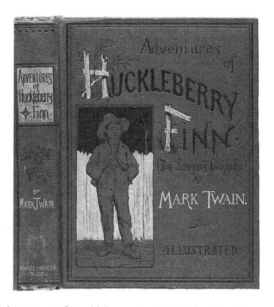

Adventures of Huckleberry Finn (1885) by Mark Twain

> I wandered lonely as a cloud
> That floats on high o'er vales and hills,
> When all at once I saw a crowd,
> A host, of golden daffodils.
>
> William Wordsworth (1804)

Works published before 1923, such as the photograph, book, and poem shown here, are no longer under copyright. Yet, Shel Silverstein died in 1999, perhaps swallowed by a boa constrictor, and his copyrights live on. Others would need permission from Silverstein's heirs to make copies of *Where the Sidewalk Ends*, publicly perform *The Cover of the Rolling Stone*, or publicly display the drawings from *The Giving Tree*. Life is short, copyright is long. Copyright now lasts 95 years, plus or minus a few years.

Pre-1978 works: 95-year term

Mark Twain argued for perpetual copyright. As he put it, ownership of a house does not expire, so ownership of a copyright should not. Copyright does last a long time. Just how long depends on when the work was created. In 1809, the first copyright law in the United States gave a total copyright term of 28 years. Obviously, those works are long out of copyright. Fast-forward through several iterations: the Copyright Act of 1909 gave a copyright term of 56 years if the copyright holder filed a renewal at 28 years. Many works went out of copyright after 28 years because no renewal was filed. Other works under the 1909 act started to go out of copyright around 1966, when works published in 1910 reached the 56-year limit.

Warm up your adding machine. In 1977, Congress added 19 years, giving an effective term of 75 years (28 + 28 + 19 years) for pre-1978 works. Works whose copyright had expired did not get additional time, so works published before 1923 (1978 – 56 = 1922) are out of copyright. *The Waste Land* is in the public domain, as it was published in 1922.

Tin Pan Alley (1914)   *Skrik* (1893) by Edvard Munch

The new term of 75 years meant that nothing from after 1922 need run out of copyright until 1998 (1923 + 75). As 1998 approached, works from the 1920s were headed for the public domain. Such works included early Mickey Mouse cartoons, *The Great Gatsby*, many Tin Pan Alley songs, and the evergreen "Happy Birthday to You." In 1998, Congress again added 20 years to the term of all copyrights. The term of pre-1977 copyrights is now 95 years. That means that nothing from 1923 or after will run out of copyright until after 2018 (1923 + 28 + 28 + 19 + 20). By comparison, if a patent had a 95-year term, then patents would still be in effect on inventions like insulin (1923), nylon (1935), and radar (1935)—not to mention the machine to slice bread (1932).

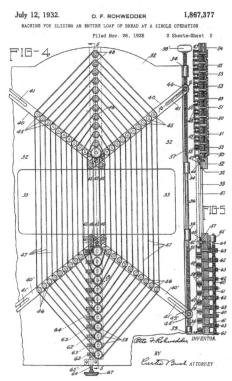

From U.S. Patent No. 1,867,377 (1932)
(Bread-Slicing Machine)

Steamboat Willie, a.k.a. Mickey Mouse (1928)

Questions

1. Here are some for Watson, the *Jeopardy*-winning artificial intelligence program. Which of these works were published before 1923 and so are not under copyright? Feel free to consult Google or Siri.[19]

    Sigmund Freud, *The Interpretation of Dreams*
    W. E. B. Du Bois, *The Souls of Black Folk*
    Franz Kafka, *The Metamorphosis*
    James Joyce, *Ulysses*
    A. A. Milne, *Winnie-the-Pooh*
    Virginia Woolf, *A Room of One's Own*
    Aldous Huxley, *Brave New World*
    Dr. Seuss, *The Cat in the Hat*

2. When would these copyrights expire?
    a. "When I'm 64" (Lennon & McCartney, 1967)
    b. "Seventy-Six Trombones" (Meredith Willson, 1957)
    c. "Bob Dylan's 115th Dream" (Bob Dylan, 1965)

3. Sometimes copyright owners use Roman numerals for the years in their copyright notices, such as "*Ben Hur*, © MCMLIX MGM." When will MGM's copyright in *Ben Hur* expire?

Answers

1. These works are no longer under copyright:
    Sigmund Freud, *The Interpretation of Dreams* (1900)
    W. E. B. Du Bois, *The Souls of Black Folk* (1903)
    Franz Kafka, *The Metamorphosis* (1915)
    James Joyce, *Ulysses* (1922)

    These works are still under copyright:
    A. A. Milne, *Winnie-the-Pooh* (1926)
    Virginia Woolf, *A Room of One's Own* (1929)
    Aldous Huxley, *Brave New World* (1932)
    Dr. Seuss, *The Cat in the Hat* (1957)

2. For pre-1978 works, the term is 95 years.
    a. 1967 + 95 = 2062 (easy way to add 95: add 100 and then subtract 5)
    b. 1957 + 95 = 2052
    c. 1965 + 95 = 2060

3. The film was published in MCMLIX, so, assuming the copyright was properly renewed XXVIII years later in MCMLXXXVII, the copyright would last for a total of XCV years (XXVIII + XXVIII + XIX + XX)—that is, until MMLIV.

    Using Roman numerals is impractical. Perhaps that is the reason some still use them. It is time for a copyright notice with the year in binary: Software Law, © 11111011110 Ian.

Post-1977 works: Life + 70, or 95 years, depending on certain factors

Monty Python and the Holy Grail, from 1975, will be copyrighted until 2070 (1975 + 95). For works created after 1977, the term is the life of the author plus 70 years. Sherman Alexie has not shuffled off this mortal coil as of 2015, so *The Absolutely True Diary of a Part-Time Indian* from 2007 will be under copyright until at least 2085 (2015 + 70). For post-1977 works by employees, the term is still 95 years, because corporations have no life to measure.

Copyright lasts a lot longer than incentives would require. No one deciding whether to write a book or even make a zillion-dollar movie considers whether it will still be under copyright 70 years after his death. As an economist would put it, the present value of possible payments seven or more decades in the future is negligible. In plain English, possible royalties 70-plus years from now will not affect the decision to write a song today. Rather, the long term of copyright better fits the idea of author's rights. To a strong believer in author's rights, if *Runny Babbit* is still selling copies in 2050, Shel Silverstein's heirs should get paid.

Questions

1. For joint authors, the term is measured by the longer life. Should authors pick young co-authors so as to get a longer term?
2. When would the following copyrights expire?
   a. "99 Problems" (Jay-Z, 2004)
   b. "99 Red Balloons" (Nena, 1982)
   c. "99 Revolutions" (Green Day, 2012)
   d. "The Pirate of Penance" (Joni Mitchell, 1968)
3. *Til death do us part, and then some.* Drudge churns out a potboiler as an employee of Roman Romances. In Drudge's spare time, he writes a poem. Which work has a longer copyright?

Answers

1. There would not be much point in such a trick. If a 70-year-old author got a 20-year-old co-author, they would get some 50 more years of copyright—but not until both were long dead. Meanwhile, the 70-year-old would have to share rights now, a big price to pay for a possible benefit to her or his heirs.
2. The answer for all four examples is, "Not anytime soon." For post-1977 works, the copyright term is life plus 70 years for individual and joint authors and 95 years for works made for hire. For joint authors the longer life is the measuring life. The authors in our examples are still going strong. The copyrights will last 70 years longer than they will. The earliest one is 1968, so that copyright has "only" the pre-1978 term of 95 years—and that is until 2053.

3. The answer for this example is, "It depends." The copyright term for works by employees is 95 years. For works by individual authors, the copyright term is life plus 70 years. If Drudge lives another 25 years, then the copyright term for the poem will exceed 95 years. A kindergartner has the copyright in her finger painting. If she lives another 90 years, the copyright will last 160 years (90 + 70).

Orphan works

General. Stop! I think I see where we are getting confused. When you said "orphan," did you mean "orphan"—a person who has lost his parents, or "often," frequently?
Pirate King. Ah! I beg pardon—I see what you mean—frequently.
General. Ah! You said "often," frequently.
Pirate King. No, only once.[20]

The Pirates of Penzance, published 1879, so long out of copyright

A side effect of the long copyright term is the "orphan works" problem. Suppose that a researcher uncovers a photo, or short story, or song in an old book. The book bears a copyright notice, "© 1940 Peirce Press." It is unclear as to who holds the copyright or whether it was renewed. Probably no one would object if the researcher published the photo/story/song/book. But a copyright holder could pop up and be entitled to considerable money for copyright infringement. A careful party would not take the risk of copyright infringement, whether on economic or ethical grounds.

Please, Sir, Could I Have Some More Copyright?

Works written but unpublished as of 1978

Mark Twain's *Adventures of Huckleberry Finn* (1885) has been out of copyright for over 100 years. But Twain's autobiography, written from 1870–1899 and published in 2000, is under copyright. Twain kept most of his autobiography private, to avoid lawsuits and duels. To encourage publication of works unpublished as of 1978, Congress gave them copyright until 2047. This was not perpetual copyright, but it would extend more than a century beyond Twain's life.

Whither copyright: Constitutional limits on copyright?

The Supreme Court Did Not Sink Copyright's Term Extension

In 1998, Congress retroactively added 20 years to the copyright term, disappointing those who expected works from the 1920s to start going out of copyright. Some argued a retroactive copyright extension was beyond the constitutional power of Congress. If the idea was to give copyrights to encourage the creation of works, it was not a valid idea. Obviously, copyright extension in 1998 did not make authors back in the 1920s work harder. They also argued that Congress went beyond its power to grant copyrights for "limited Times" by repeatedly extending copyright terms, called by some "perpetual copyright on the installment plan." But the Supreme Court upheld the law, ruling that Congress has great scope in fashioning copyright law. Copyright law itself was deemed to protect freedom of expression, because it permits fair use and allows copying of ideas.

The issue may arise again in 2018.

What Is Copyrighted: A Creative Work of Authorship in Tangible Form

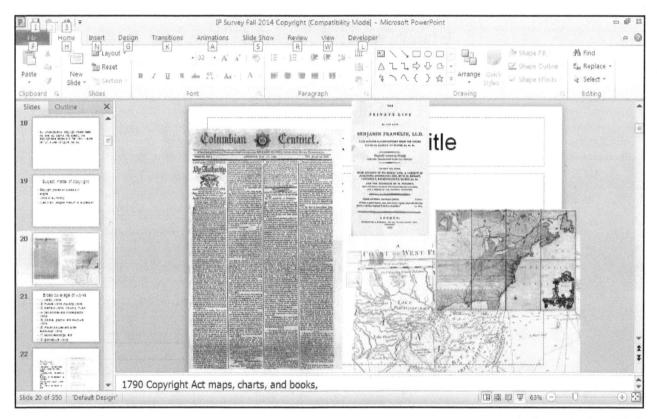

An Audiovisual Work

The art of Biography
Is different from Geography.
Geography is about maps,
But Biography is about chaps.

Edmund Clerihew Bentley

Almost all works of geography and biography were covered by early U.S. copyright law.[1] The first U.S. copyright law, in 1790, covered books, maps, and charts. Charts are maps for sailors, so the 1790 copyright really covered just books and maps. Congress has steadily broadened copyright over the years, making it potentially apply to any creative work of humans.

Works of authorship

Copyright applies to any creative *work of authorship* in a tangible form.[2] Works of authorship include the following categories: literary works; musical works; dramatic works; pantomimes and choreographic works; pictorial, graphic, and sculptural works; motion pictures and other audiovisual works; sound recordings; and architectural works.[3]

Literary works

Following are snippets from two literary works:

> Should Heaven send me any son,
> I hope he's not like Tennyson . . .
>
> Dorothy Parker, *Alfred, Lord Tennyson*

magicTools = (new ToolMaterial(1, 208, 6.5F, 1));
magicAxe = (new ItemAxe(magicAxeID, magicTools)).
setItemName("magicAxe"); magicHoe = (new ItemHoe
(magicHoeID, magicTools)).setItemName("magicHoe");

> Notch, source code for *Minecraft*

Literary works include works using words, letters, numbers, or other similar symbols, Literary works go beyond literary forms like novels (*The Great Gatsby, Harry Potter og viskusteinninn*), short stories (*Birthmates, The Library of Babel, The Secret Life of Walter Mitty*), and poetry (*We Real Cool, Howl, The Wasteland*). Literary works include the software code for the Android operating system, the warning label on a bottle of sleeping pills, e-mail messages, love letters, the text of Dr. Martin Luther King's *I Have a Dream* speech, a movie script, and an article in *Science*.

Musical works

Three Musicians (1921) by Picasso

Musical works of all genres may be copyrighted, from "Rhapsody in Blue" to "Rep Yo City," from a Prokofiev sonata to a Sinatra song. Some musical works benefit from the full 95 years of copyright. Tin Pan Alley songs from the 1920s such as "Yes, We Have No Bananas" (1923) and "Happy Days Are Here Again" (1929) are still bringing in licensing fees. "Happy Birthday to You" (registered 1935, although some say it was written before 1923 and so should be out of copyright) has a copyright that brings in millions. "Over the Rainbow" (1939) and "Your Cheatin' Heart" (1953) are still going strong. Musical works play a big role in debates on copyright policy. Some attribute a decrease in music sales to unauthorized sharing.

Dramatic works

Shakespeare, Published before 1923, Is No Longer under Copyright

Dramatic works are ones in which the action is acted out, or performed, as in a play or an opera. The book *Harry Potter* is not a dramatic work, even though it includes drama, but the *Harry Potter* movies are. An opera is a musical work, a literary work, and a dramatic work.

Pantomimes and choreographic works

Dancers in the Wings by Edgar Degas

Moulin Rouge Poster by Henri de Toulouse-Lautrec

Mimes and dancers get their own category. "Choreographic" would not cover football plays or professional wrestling but is limited to dance works. A football team's playbook would fall into the next category.

Pictorial, graphic, and sculptural works[4]

Skrik by Edvard Munch (1893)

Pictorial, graphic, and sculptural works can include <u>any type of artwork in two dimensions </u>(paintings, drawings, blueprints, photos, doodles) <u>or three dimensions</u> (sculpture, furniture, models, Lego villages).

Motion pictures and other audiovisual works

The Kid (1921)

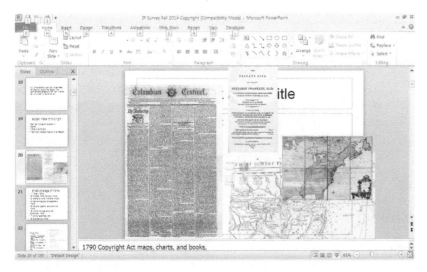

Example of an Audiovisual Work without Audio

We know movies are copyrighted from the FBI warning that starts each film. An <u>audiovisual work</u> is a series of <u>related images shown by a device, perhaps with sound</u>. Audiovisual works include video games, slide shows—whether film or PowerPoint—and film strips of yore.

Sound recordings

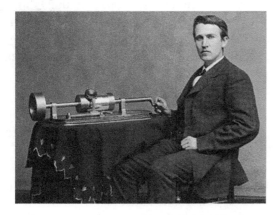

Thomas Edison and Sound Recording Device

Sound recordings of all genres may be copyrighted, from "Rhapsody in Blue" to "Rep Yo City," from "Moonlight Sonata" to a song by Frank Sinatra. Sound familiar, as it were? A <u>recording often embodies two copyrighted work</u>s: th<u>e musical work</u> <u>and</u> the <u>sound recording</u> made of someone performing the music. <u>One party may hold both copyright</u>s, such as when a music company, by contract, owns the copyright in both the song and the recording of the song. <u>But often they are held by separate parties.</u>

Questions

1. If a movie producer wants to get permission to use a hit song in the soundtrack, whom should she contact?
2. Audubon records a nightingale singing. Is this copyrighted?

1. She should contact the copyright owner(s). There will be one copyright for the song, one for the sound recording. Both may require payment—which is why you often hear cover versions of songs in movies and commercials. It may save licensing fees to pay the music copyright holder and rerecord the song.
2. Yes—not in a musical work (the nightingale's song), but in the sound recording.

Architectural works

Congress added the category of architectural works to the Copyright Law in 1990 as part of the United States joining the most important international copyright treaty (more about this later). Congress also included a limit on the copyright: anyone may make a picture of a building that is visible from a public place.

Other works of authorship

What about a work that does not fall into any of the foregoing categories? Those categories are so comprehensive that few things made by people do not fall into any of them. Works directed at smell or touch are not specifically mentioned, although a work in Braille would likely be a literary work. Whether a symphony of squeezes or smells is copyrighted may someday pose a puzzle for the courts.

Question

Into which categories would each of the following fall?
Life of Pi, a novel
Life of Pi, a movie
"Pi Symphony" (assigned notes to the digits of pi)
Recording of a lecture about pi
The software code that runs the Raspberry Pi
The Globe Theatre—and the plays performed there
"Deuce Coupe," a dance

Answer

Life of Pi (novel)—literary work
Life of Pi (movie)—motion picture and other audiovisual works
"Pi Symphony" (assigned notes to the digits of pi)—musical work
Recording of a lecture about pi—sound recording
The software code that runs the Raspberry Pi—literary(!) work
The Globe Theatre—and the plays performed there—architectural work, literary works, dramatic works
"Deuce Coupe" (dance)—pantomimes and choreographic works

Not a work of authorship

Copyright applies only to works of authorship. Lady Gaga and Stephen Colbert have creatively developed distinctive personas, but those, as such, are not copyrighted works. Likewise, anyone may copy bon mots and escapades. Works by nonhumans are also not copyrighted. Paintings by elephants or videos by camera-stealing seagulls are not works of authorship. A natural garden created by creeping vines and wandering seeds is not the work of an author, for copyright purposes. Theology sees it differently.

A mere word or phrase is not a work. If someone could copyright a word, then others could infringe the copyright simply by using the word. Words and phrases may be trademarks, however. Hence the common misreport that someone has "copyrighted" a word or phrase. But trademarks may be freely used as words by others. Thus, you can talk about Apple all you want without infringing copyright. But putting the name "Apple" on products you sell infringes trademark.

Questions

1. On an online forum, Boingboing asks, "What is the purpose of a zero Ohm resistor?"[5] The comments:
   > Resistance is futile. . . .
   > I bow in ohmage to you. . . .
   > Watt? . . .
   > These puns are revolting. . . .
   > It's the current trend. . . .
   > I'm galvanized into action. . . .
   > Power to the people! . . .
   > Wire y'all doing this? . . .
   > Oh god . . . don't divide it by zero. . . .
   > STAY AWAY FROM OHM'S LAW!!! . . .
   > I believe they are used in electric left-handed bacon stretchers. . . .

   Are the comments works of authorship?
2. "Intellectual property" is an imprecise, somewhat misleading, catch-all term for copyrights, patents, trademarks, trade secrets, rights of publicity, and other intangible legal rights.[6] These things are not necessarily intellectual or property. If you came up with a better term, could you copyright it?

Answers

1. Many are short phrases, which are not copyrightable.[7] As to the longer ones, a court might still treat them as ordinary conversation, which is not a work of authorship. Even if the longer ones might be judged "works of authorship," each could be freely copied under the principles of fair use and the nonprotection of ideas, as discussed later in this book. (Bonus: The first one quotes from *The Hitchhiker's Guide to the Galaxy*, and so is clever but not creative—see the discussion of creativity later in this chapter.)
2. No copyright exists in words or short phrases. A really long term could be copyrighted, but the length would defeat the purpose.

Artificial intelligence may become creative,[8] but for the last 40 years, that date has been 10 years in the future. Software like Watson and Deep Thought, which can beat humans at *Jeopardy!* and chess, make artificial human intelligence seem close. Software like AutoCorrect, not so much. As the *New York Times* reported, Stephen Fry tweeted, "Just typed 'better than hanging around the house rating bisexuals' to a friend. Thanks, autocorrect. Meant 'eating biscuits.'" The *Times* also noted changes like "geocentric" to "egocentric," "catalogues" to "fatalities," and "Iditarod" to "radiator," and asked, "What is the logic?"[9]

Question

Photo produced with equipment of David Slater

A noted nature photographer leaves his camera unattended on a tripod. A macaque monkeys around with it, inadvertently making a self-portrait. Copyrighted?

Answer

If the photographer simply was unmindful, then no human author = no copyright. But if the photographer planned things that way, he would be the author.[10]

Government works are not copyrighted

Migrant Mother, by Dorothea Lange, U.S. Farm Security Administration (1936)

A Sonic Boom, Ensign John Gay, *USS Constellation*, US Navy, The Astronomy Picture of the Day

apod.nasa.gov

If someone sells copies of the Copyright Act, she is not infringing a copyright in the act itself. That literary work states, self-referentially, that <u>copyright does not apply to works of government employees.</u> NASA produces stunning, copyright-free images. Government photographers have likewise produced valuable images, such as the "Situation Room" shot, showing the U.S. president and his advisers watching the operation seeking Osama bin Laden. Because the photo was not copyrighted, it could be freely redistributed—and adapted. In one version, all those present are dressed as superheroes.

"Situation Room," Pete Souza, White House

Lack of copyright does not mean that all government information may be accessed. The photo was released with this proviso: "Please note: a classified document seen in this photograph has been obscured."

Government reports are copyright free and so may be freely resold or adapted. Some are a little dry, admittedly, such as the USDA's "Cotton Ginnings: Running Bales Ginned by Crop, Excluding Linters, 2009–2012." By contrast, an investigative report by the U.S. Navy detailed how the commander of a nuclear submarine officer faked his own death to end a romantic affair.[11] A possible musical? *Not the Very Model of a Modern Major General*, asking who has a finger on the red button?

Government Legal Documents Are Not Copyrighted

Similarly, copyright does not apply to most documents produced by federal, state, tribal, or local governments. Government needs no incentive to produce works. More importantly, a government should not be able to prevent people from copying and distributing its works. If an embarrassing memo leaks from the White House, the government cannot suppress publication by claiming the right to make and distribute copies. Much valuable material is not copyrighted, such as regulatory reports, judicial opinions, and briefs filed by government lawyers.

By contrast, the United Kingdom recognizes Crown Copyright if "a work is made by Her Majesty or by an officer or servant of the Crown in the course of his duties." Many such works, however, are freely released under an Open Government License.

Works created by nongovernment employees are copyrighted even if funded by the government. Statues for official monuments are often created by artists on commission but not employed by the government. Such statues are copyrighted, but fair use would allow you to take a picture of them.

Questions

1. The U.S. Government Accountability Office's staff writes a report, *Actions Needed to Evaluate the Impact of Efforts to Estimate Costs of Reports and Studies*. The report wins the 2012 IgNobel Prize for Literature: "The U.S. Government General Accountability Office, for issuing a report about reports about reports that recommends the preparation of a report about the report about reports about reports." Is the prize-winning report copyrighted?
2. The Smithsonian American Art Museum owns paintings by artists from Georgia O'Keeffe to Roy Lichtenstein. Are they copyrighted?

Answers

1. The report is a government work, thus it is uncopyrighted. Anyone may make copies of the report, make sequels, distribute copies, display it to the public, or perform it in public. Not many will.
2. If O'Keeffe or Lichtenstein had painted the works while on the job for the government, the paintings would be uncopyrighted government works. As it happens, Lichtenstein did work as an artist and engineering draftsman in the U.S. Army when young, but those works are not in the museum. The paintings were painted by private citizens, then purchased by the government. (Question: who owns the copyright? The artist, unless he sold it along with the painting. Ownership of a physical work and ownership of the copyright are separate.)

Creativity

Copyright applies only to a *creative* work of authorship in a tangible form.

Plenty of valuable material cannot be copyrighted: a business treasures its customer list; a security camera video solves a crime or immortalizes a shopper knocking over a pyramid of soup cans; a university's grade records are important to the school, the students, and prospective employers; scientific data from experiments represents huge investments in time and other resources; MRI images save lives; a 3D virtual copy of a car made with robotically guided precision lasers can be used in engineering, not to mention marketing. None of these is copyrighted.

Copyright applies only to *creative* works of authorship. The preceding works all lack creativity. Facts are discovered, not created, so the items on the customer list and data from experiments are not copyrighted. There was no creativity in selecting the customer list (it simply lists all the customers in order); the grade records (likewise listing all the grades reported, linked to the relevant student record); the security video (the camera was mounted in standard fashion behind the cashier and set to run 24/7); the scientific data (although gathered skillfully, it simply reported facts); the MRI (which imaged the relevant area according to protocol); or the 3D virtual copy of the car (made by an automated process).

There could be creativity in such works. *The Onion* could write a comical, fictional list of customers. A moviemaker could film a scripted scene from the perspective of a security camera. One could make a funny MRI. One could artistically draw up a 3D image of a concept car. But without an element of creativity, there is no copyright.

Gathering facts may require resources, whether the facts are valuable for business, like a customer list, or research, like scientific data. Making exact copies may require considerable technical skill and fancy equipment, like the 3D virtual copy. But copyright applies only if some creativity has been used.

The low requirement of creativity is easily met: hack novels, formulaic songs, academic drivel, predictable genre movies, and fatuous political commentary are all copyrighted.

Note: If an author wrote a book about a famous crime but made up a lot of the "facts," he could not recover for infringement if others copied those fictitious facts. Facts are not protected by copyright, so others could copy them freely, even though the book is a copyrighted work.

Questions

1. May a newsgatherer copy facts, sports scores, stock prices, and other facts from newspapers and other media without infringing copyright?
2. Is the *New York Times* list of best-selling books copyrighted?
3. A court reporter uses skill to get a perfect verbatim trial transcript. Copyrighted?

Does a Stenographer Get a Copyright?

4. Amazon.com offers the AudioQuest K2 terminated speaker cable for $8,450. Yes, an $8,450 speaker cable. A snippet from the "Most Helpful Customer Review":

   > We were fools, fools to develop such a thing! Sound was never meant to be this clear, this pure, this . . . accurate. . . . We live underground. We speak with our hands. We wear the earplugs all our lives.[12]

   Is the review copyrighted? Bonus question from a coming chapter: If so, who owns the copyright?

5. Is a map copyrighted? It just shows facts, after all.

Answers

1. Yes. Facts are not copyrighted even if they are valuable.
2. Although great efforts go into compiling the list, it represents facts. There is no creativity in selecting, arranging, or coordinating the list (or if there is, the *New York Times* has some explaining to do). Organizations that depend on selling information (from stock data to addresses to weather information to astronomical data) often find the lack of copyright protection irksome.
3. The court reporter captures only noncreative material, the exact words spoken by others. No copyright.
4. More than enough creativity here to qualify for copyright. As we will see later, the copyright belongs to the author (although Amazon's terms of service may transfer or grant license to use it, if the author agrees).
5. Maps provide information, but also reflect many creative aspects: what to include, how to portray features, how to arrange things. The factual information is not copyrighted, but the map itself is.

Copyright Litigants Battled about This Widely Reprinted Photo of Oscar Wilde

Technology and creativity

Does a photograph merely record facts in the world, or does it reflect copyrightable creative expression? A purely mechanical process does not involve creativity. But a work made using technology can be creative. A photographer makes plenty of creative choices: framing, angles, lighting, poses, props. Likewise for other works made with devices: film, video, sound recordings. Data gathered by devices that simply record facts, by contrast, would likely not be copyrightable; cardiographs, keystroke recording software, meteorological instruments, stock tickers.

Security Camera Still: Creative?

Question

Could the printout from a seismograph be copyrighted?

Normally, a seismograph's recordings of vibrations would simply be noncopyrighted facts. But if Jinsky placed a seismograph and leaped around to get some curves, that would supply the necessary creativity.

Adaptations and creativity

The musical *Rent* is copyrighted, even though it adapts the opera *La Bohème* (published 1896, so no longer under copyright—and itself based on a novel). Any number of adaptations of Shakespeare's works are copyrighted. *Romeo and Juliet*'s offspring include *Abie's Irish Rose*, *West Side Story*, *Gnomeo and Juliet*, and *The Lion King II*, not to mention *Romeo and Juliet* in film, opera, and ballet. The 1929 film of *The Taming of the Shrew* had "the ultimate in screen credits: 'By William Shakespeare, with additional dialogue by Sam Taylor.'"[13] An <u>adaptation that adds creative expression to a public domain work will be copyrighted</u>. So works that are not under copyright may be used as raw material for works that will be copyrighted. A work need not spring full grown from the author's head. *The Hunger Games* and the *Percy Jackson and the Olympians* series borrowed heavily from Greek myths but are certainly creative enough to qualify for copyright.

An adaptation may simply add creative expression to the original. An annotated edition, a translation, a Bowdlerization, or even a shortened version (creativity can come in excising) will be copyrighted provided a little creativity is used. Even a new translation of the *Odyssey* or the Bible will be copyrighted.

Question

Is there copyright beyond the grave? Jane Austen's *Pride and Prejudice* (1813) is long out of copyright. *Pride and Prejudice and Zombies* (2009) adds considerable ultraviolent zombie mayhem to the original story. Copyrighted?

Answer

Yes, an adaptation is copyrighted even if it uses preexisting material. A theoretical question: What is the copyright term for a work authored by a zombie? The normal term of life plus 70 years does not work so well for the undead.

The flip side: Copyright protects only creative expression

An author can add a little creativity to preexisting stuff and qualify for copyright. A list of odd place names culled from maps, *Romeo and Juliet* transposed to 2013 Chicago, a photo of the *Mona Lisa* made with a little creativity, a compilation of medical data creatively selected, coordinated, or arranged . . . all qualify for copyright. But copyright protects only the new creative elements. Copyright does not protect noncreative material, such as facts or material created by someone else. Some might infringe copyright by copying the entire list of odd place names, because that would include copying the author's creative selection and arrangement, but not by copying some of the names, which would not copy the selection or arrangement. Likewise, someone could freely copy some facts from a database or book. If she copied the entire database or book, that might be infringement for copying the selection and arrangement of the facts. Works that are largely made of noncreative material have "thin" copyright, meaning that a lot of stuff may be copied without infringing.

Questions

1. *Life Plus 700?* is a book of early music that collects works of such composers as Palestrina, Monteverdi, and Josquin—none later than 1643. But it bears a 2013 copyright notice. Can the book be copyrighted? If so, which parts are copyrighted?
2. Photographer Harney snaps socialite Clark Rockefeller and his daughter leaving a Boston church. Rockefeller turns out to be an imposter from Germany, Christian Karl Gerhartsreiter. Police publicize the photo in their search for Gerhartsreiter. A movie about him copies the photo. Copyright infringement?

Photograph of Clark Rockefeller

Movie Still Showing Recreation of Photograph of Rockefeller

Answers

1. The book could have copyright as a compilation if it featured the necessary creativity in the selection and arrangement of the material. Or it could have copyright as an adaptation if it included creativity in musical arrangement, transcription, simply filling in blanks, or annotations. What parts are copyrighted? The new,

creative elements added by the authors. But here's the rub: you can't tell by simply looking at the book what the authors have added. As with many works containing copyrighted elements and noncopyrighted elements, it is not at all clear which are which.

2. The movie did not copy creative expression from the photo, but only facts—that is, the child sitting on the imposter's shoulders. There is no infringement. Had Harney posed the couple, the result would be different.

Creativity in tangible form

Tangible Form of a Dance: *ROADRUNNERS* (1979) by Merce Cunningham

Copyright applies to a creative work of authorship *in a tangible form.*

As soon as a paparazzo snaps a photo of a celebrity, the photo is copyrighted. He need not register the copyright, or put © on a copy, or sell a copy, or even show it to anyone. Copyright attaches when the work is put in tangible form. Copyright is more user friendly than patent, which has a demanding application process. Authors get copyright without even thinking of it, simply by writing, painting, sculpting, or recording.

Some artistic forms do not inherently put the work in tangible form. A dance may leave no trace.[14] Jazz, improvisation, comedy, performance, and conceptual art are intangible. But a ballet could be recorded in choreographic notation, described in verbal instructions, videotaped, or recorded in instructions in Morse code. A stand-up comic who wants copyright in her routine can simply write it down or videotape it.

NASA/JPL-Caltech/Mars Science Laboratory, Astronomy Picture of the Day, http://apod.nasa.gov/apod/

Is there such a thing as a Martian copyright? Would a photo of Mars by Curiosity be copyrighted under U.S. law?

Answer

This question raises several issues. If the picture was made auto-matically by Curiosity, then it would have no human author. Some-one having written the software that eventually took the picture would not be enough for copyright. If someone back on Earth made creative decisions, say, maneuvered the camera and snapped the shot, subject to a 14-minute delay even at the speed of light, then that could be a work of human authorship with the necessary tiny bit of creativity. If it were a NASA employee, it would be U.S. government work, and so not subject to copyright. If it were controlled by an independent contractor or a kid who hacked into NASA's network, then it could be copyrighted.

Movie Rights, Book Deals, Recording Contracts, Free Software: Copyright Transactions

A woman must have money and a room of her own if she is to write fiction.

Virginia Woolf

You know I'd go back there tomorrow
But for the work I've taken on
Stoking the star maker machinery
Behind the popular song.

Joni Mitchell, "A Free Man in Paris"

Who gets the copyright?

Shakes, a playwright, works with a dramaturge to hone his play, *Lear Jet*. Edison, an employee of Othello Software, designs and codes a new telephone operating system. Hameron, a film director, micromanages the making of *Gigundous*, although she pays others to write the script, film the action, and edit the movie. Who owns the copyrights?

If the parties have been careful, the answer lies in their contracts. Where more than one party is involved in the creation of a work, they can agree as to who will hold the copyright. They can split the pie, or agree that one gets the lion's share. A contract could state that Shakes owns the copyright but the dramaturge gets 31.41 percent of all royalties from the play and any adaptations (movies, sequels, Icelandic translations). Edison and her employer could agree that the employer gets the copyright but Edison gets a guaranteed salary for 3.14 years.

The producers of *Gigundous* might get the copyright by contract in return for agreed payments to Hameron, the writers, and the crew. Perhaps Hameron would get a percentage of the gross as well.

Even where money is not important, agreeing about rights can clarify things. If a nonprofit theater group writes a play, agreeing about the rights at the outset reduces the chance of later conflict concerning who can write the sequel, or whether to give permission to another troupe to perform the play, or even, happily, how to split the money if somehow it becomes a commercial success.

Many times the parties do not agree in writing about who will own the copyright. They may not even think about the matter. A family business hiring a designer for its logo, some teenagers writing songs in a garage, some teenagers putting together a Linux disco (*distro*, confound you, AutoCorrect) in a garage, or a couple hiring a photographer for their wedding may not have copyright on their minds. Even when parties think of copyright, they may not reach an agreement about assignment of rights. They may have more pressing issues. Rights can be a delicate issue: they may simply rely on the law to assign rights. Many do not even realize that they can decide the matter, as opposed to copyright law mandating whom the copyright belongs to.

Increasingly, parties do address copyright issues in their contracts. Even employees with the most mundane jobs may have a clause in their contract assigning to the employer all copyrights, patents, trademarks, trade secrets, and rights of publicity. Websites often address copyright in the terms of service, sometimes overreaching. Universities more and more often have copyright and patent policies that apply to both students and staff—and they vary quite widely as to who gets what rights. Agreements for joint ventures, investments, art commissions, and research grants now often address ownership of copyright, patent, trademark, or trade secrets springing from the work.[1]

But not always. If the parties do not decide who owns the rights, copyright law fills the gap. The default settings are that individual authors own the copyright to their works, joint authors share copyrights, and employers own the copyrights to employees' works.

One author

The author of a work owns the copyright, until she sells it, gives it away, signs it away, or goes into bankruptcy. Suppose Herzog makes a documentary of pedestrians walking on a busy city street. Many people will appear in that documentary, but Herzog will be the author of the video. Even if a crew of gaffers, camera people, and sound technicians perform the technical work, the filmmaker will be the author if she controlled the creative expression.

Question

> Recall that Amazon.com customer review ("We were fools, fools to develop such a thing!")? Suppose Amazon decides to include it in a book—or adapt it into a book, movie, or Broadway musical. Who owns the copyright in the customer review?

Answer

> The author owns the copyright—unless otherwise agreed. Whenever ownership or other rights are at issue, we must check to see

if they are governed by agreement. Under Amazon.com's terms of service (often called TOS), "If you do post content or submit material, and unless we indicate otherwise, you grant Amazon a nonexclusive, royalty-free, perpetual, irrevocable, and fully sublicensable right to use, reproduce, modify, adapt, publish, translate, create adaptations from, distribute, and display such content throughout the world in any media." If the TOS is effective, the author owns the copyright, but Amazon can use it any way it likes.

Joint authors

If twin brothers Thomson and Thompson co-author a book, they are joint authors. A publisher, moviemaker, or sequel writer would need permission from only one or the other. As joint authors, each twin would have the right to authorize use of the work, but he would be obliged to split profits with the other. A joint author has only limited control over the fate of the work. If there is little commercial value (such as in an academic paper), then the obligation to share profits won't matter much.

Not everyone who contributes to making a work gets the status of joint author. The musical *Rent* needed some fixing up before its off-Broadway production. The author, Jonathan Larson, worked with a dramaturge, Lynn Thomson, to clarify the music and storyline. Larson died shortly before the musical opened and went on to be a smash. Thomson sued Larson's estate for a share of the royalties. But the parties had not intended to give her joint author status. All billing, contracts, and promotion portrayed Larson as the sole author. Larson had complete control over the script. Copyright law says to the dramaturges, editors, improvisers, and script doctors of the world, "If you want author status, get agreement from the principal author." Giving joint author status to every contributor would create a tangled web of ownership. After all, actors improvise lines, editors fix stories, and camera operators put new perspectives on films.

Questions

1. In what categories of works are joint authors common? Uncommon?
2. Giacomo Puccini wrote the 1896 opera, *La Bohème*, on which *Rent* was based. Was Puccini a joint author of *Rent*?

La Bohème Poster by Adolpho Hohenstein (1896)

1. Songs often have joint authors: Drake & Lil Wayne, Lennon & McCartney, Carole King & Gerry Goffin, Elton John & Bernie Taupin, Gilbert & Sullivan. Symphonies less so. Beethoven and Mahler were sole authors, although this does not mean that they got no help. But contributing help, as explained earlier, is different from co-authoring. Many literary works have joint authors, including textbooks and academic papers. Scientific papers often have more than a dozen authors. Celebrity autobiographies often have an author and a ghostwriter, with or without credit. But, apart from *Good Omens* by Gaiman and Pratchett, no great novel, short story, or poem had more than one author. *The Autobiography of Alice B. Toklas* had one author, Gertrude Stein. Paintings and sculpture typically are the creative work of one artist. Many artworks are credited to one artist, even though the physical work may have been done by anonymous employees, from Warhol's Factory to Koons's corps of assistants.

2. Puccini is not a joint author of *Rent*, simply because Larson did not intend Puccini to be treated as a joint author. Otherwise the copyright term of works would be infinite. Shakespeare's heirs would be due a lot of royalties.

Works by employees

> The fault, dear Brutus, is not in our stars,
> But in ourselves, that we are underlings.
>
> Cassius, in Shakespeare's *Julius Caesar*

Much of intellectual property law is employment law. Employees create copyrightable works, patentable inventions, and valuable trade secrets. Who owns that intellectual property is often in dispute.

The copyright in a work by an employee on the job belongs to the employer, unless otherwise agreed—and few employees have that much bargaining power. The category of "work made for hire" is not as broad as it sounds. Simply because a party pays for a work to be made does not make it a work made for hire. A work made for hire is one created by an employee within the scope of her employment. Software written by engineers, stories written by journalists, and skits written by television writers may all be works made for hire if the author was working as an employee. But software written by consultants, stories written by freelance writers, and skits written by playwrights with grants are not works made for hire. The copyright does not belong to the hiring party in those cases unless the hiring party got an assignment of the copyright in a written contract.

This is a trap for the unwary. Suppose Mr. Plow, a local snow-plowing company, hires independent contractor Frink to write software that handles Mr. Plow's scheduling, billing, and client relations. Unless Frink agrees that it will belong to Mr. Plow, the copyright will belong to the author, Frink, not to the party that funded it, Mr. Plow. When a contractor is hired to produce software, literature, a movie, or some other work, the parties should negotiate and agree in writing about who will have rights to the work. They may agree that the hiring party will own the copyright. For some categories (such as translations, contributions to movies, or textbooks), they can even agree that the work will be a work made for hire.

Questions

1. *Intern's revenge:* Asok designs and codes a game, *Minimum Wage*, as an intern at ExploCorp. Asok does the work all at ExploCorp, using its equipment. The game is a smash hit. Who owns the copyright?
2. Who owns the copyright in articles written by freelancers for *The New Yorker* magazine?

Answers

1. Asok is not an employee, so Asok owns the copyright *unless* Asok signed an agreement assigning copyright in his work to ExploCorp.
2. Freelancers are independent contractors, not employees, so they would own the copyrights *unless* they assigned their rights to *The New Yorker*. Reportedly, after a string of best-selling books and movies adapted by freelancers from such articles, *The New Yorker* requires freelancers to sign over the copyright. Fair?

Works done on the job?

Scott Adams wrote his *Dilbert* cartoons in the evenings, at home, while he was working as an engineer at Pacific Bell. He was employed as an engineer, not a cartoonist. They were not works done by an employee on the job. Adams, not Pacific Bell, owned the copyrights.

Some journalists took jobs at a slaughterhouse and surreptitiously made videos showing unsanitary practices. Trying to suppress the videos, the slaughterhouse had the chutzpah to claim ownership of the copyright, but making the videos was not part of the journalists' jobs at the slaughterhouse. The copyright therefore belonged to the journalists' other employer, a television network.

Employment contracts

An employer automatically gets the copyrights in employee works. But employers frequently have employees sign contracts assigning rights to the employer anyway. There are several reasons for this. First, it makes things clear. If the employee reads the contract a (big "if"), she will see that she has signed away various rights. It will cover not just copyright, but rights to inventions, ideas, and other information. It may even go beyond the job. Some contracts apply to works done by the employee at home or works that go beyond the employee's assigned duties. But courts may not enforce clauses that go too far in this respect.

1. *Will work for IP:* Ars Technica reports that some developers leave their jobs at video game studios so they can own the games they develop at home in their spare time.[2] Why don't the games belong to the developer?
2. What if a clause in the contract assigned ownership to the studio of any game the developer ever created, even long after the developer left employment at the studio?

Answers

1. Games made in the employees' spare time and not part of their assigned work would not automatically belong to the employer. The developers must have agreed to give to the studios the copyright in any games the developers would create, whether in the scope of their job or not. IP rights (or not, as here) may be part of the employment package, just like salary and health benefits.
2. Such clauses in employment contracts are not effective if too strict. The courts may deem assignment of a life's future work to be going too far.

Copyright transactions

A car owner can use many transactions to get value from it. Cars are sold, leased, rented, consigned, hired with drivers, auctioned at Sotheby's or on eBay, raffled, offered as competition prizes, loaned to museums or to friends, bequeathed, and hypothecated, the latter being a fancy word for "used as collateral for a loan."

Ditto for copyrights. When J. K. Rowling wrote the *Harry Potter* books, she gained the rights, among other things, to make and sell copies of the books, make and show the movies, and sell *Harry Potter og viskusteinninn*. Rowling's specialty is writing, not publishing, cinematography, or translating *Harry Potter* into dozens of languages. She could use copyright transactions to exploit her various rights: that is, to sell the publishing rights (hardcover, paperback, Kindle, audiobooks, domestic and foreign), the movie rights, and even the rights to make a Broadway musical if she so chose. Copyright transactions allow an author to exploit the work commercially.

Copyright transactions go beyond commercial considerations. One transaction is simply to allow others to make uses that fall within the copyright. An author gives permission for free for various uses of her work. A writer consents that a short story be included in an anthology, a photographer agrees that a photo could be used in a political campaign, a scientist agrees that a hefty portion of an article be included in a textbook, a playwright agrees that her play could be performed by a school.

An author can also use her copyright to control the artistic fate of her work. If Rowling does not wish to see *Harry Potter* on Broadway, she can decline to sell those rights. If she wishes to control how *Harry Potter* is portrayed in films, she can retain control in the contract (including working on the screenplay herself, as did the author of *The Hunger Games*). A playwright agrees that a theater troupe can perform his play, but he may include conditions such as forbidding changes to the script or other alterations. Samuel Beckett (*Waiting for Godot*) preferred to translate his work from French to English (and vice versa) himself. He (and now his estate) licensed performances only under strict conditions to stick to the script and stage directions. Edward Albee (*Who's Afraid of Virginia Woolf*) licensed his plays for performances on the condition that the theater would not be segregated by race.[3]

The ownership rules above can all be altered by contract. An employee with negotiating power or skills may get the copyright in her work by agreement. An independent contractor may, if the hiring party is savvy, sign over the copyright in the software/graphic design/ghostwritten autobiography she is paid to create. A playwright may agree to give the dramaturge an equal share of the copyright.

Exclusive versus nonexclusive grants

A basic distinction between transactions is whether a grant of rights is exclusive or nonexclusive. An exclusive grant *excludes* a second grant of similar rights. If J. K. Rowling sells Scholastic the exclusive right to publish *Harry Potter* in hardcover in the United States, then Rowling could not authorize Penguin Books to publish a U.S. hardcover edition of *Harry Potter*. A publisher or moviemaker will normally bargain for exclusive rights.

A nonexclusive grant allows the copyright holder to grant similar rights to others. By contrast, if Ian buys a Windows 8 phone, Microsoft will give him a license to use Windows 8 on a phone. But it will not be an exclusive license—Microsoft will be free to allow others to use Windows 8 on other phones. An exclusive license may be narrow: J. K. Rowling could grant the exclusive rights to print a paperback version of *Harry Potter og viskusteinninn*, which would leave her free to grant rights to a hardcover version, an audiobook, a Kindle edition, or a paperback in any other language.

Copyright law has one formality requirement for transactions. Any transaction other than a nonexclusive license is effective only as a signed writing. If J. K. Rowling says to Warner Brothers, "You have the exclusive right to make a *Harry Potter* movie," that verbal statement would not be effective. To get an effective deal, one must put it in writing. The writing requirement gives good evidence of the grant, encourages the parties to make clear their terms, and gives the copyright owner a chance to think twice before parting with rights.

Questions

1. Sony Pictures Entertainment reportedly paid $3 million for the film rights to the biography of Steve Jobs.[4] Facts and ideas are not copyrighted, so Sony could have freely used all the facts from the book in making a movie. Why pay?
2. Should the user manual for the iPhone be considered a legally binding contract?
3. Old Mr. Burns passes away. Who now owns his vast song collection purchased on iTunes?

4. Bryant's employment contract grants Mattel ownership of his "works and inventions." Bryant has an idea for Bratz dolls. Bryant leaves Mattel and takes the idea to a new employer, who makes a fortune. Does Mattel own the idea for Bratz?
5. A publisher offers to buy the copyright to graduate student Emma's novel. What terms should she consider?

Answers

1. This example serves as another reminder that copyright is only part of the picture. Sony could indeed use the facts freely. But the deal gives other advantages. Sony can copy the way the writer expressed those facts: the creative narrative and description and the selection and arrangement of facts. Sony can also enlist the writer in thinking through how to translate the book into movie form. In addition, it will keep the writer from working on a rival project. For these reasons and more, studios often pay for movie rights to factual works. Not that they need permission to make the movie or use the facts, but they can use the expressive elements and get services and other information.
2. The user's manual will not be a contract unless the manufacturer is foolish enough to make it so. Nor should it. Otherwise, it would have to be written as a wordy legal document, full of disclaimers, as opposed to a useful guide.
3. The first place to look for an answer to this question is at any relevant contract. Here, the iTunes terms of use, like many, state that licenses terminate on death, so the songs do not go to Burns's heirs. Whereof: "How is it that copyright lasts 70 years after death, but licenses expire at death?"[5]
4. No. A contract could assign ownership of ideas, but this one applied only to concrete things: works and inventions.
5. A few things to think about:
   a. What rights should she sell? Just the rights to publish the novel, or the novel and sequels, or the novel and movie rights? If she sells the entire copyright, the publisher gets all those rights.
   b. Royalties: a flat price or a percentage of sales?
   c. Termination rights: if the book goes out of print or the publisher goes belly up, should Emma get the rights back so she can distribute her book and not leave it in limbo?
   d. A nonexclusive grant: should Emma retain the right to publish?
   e. Creative control: Emma may wish to retain veto power over edits and changes, not to mention translations.
   You can think of more. Emma should remember that everything is negotiable.

Music: The mechanical license

There is only one series of *Harry Potter* movies. Occasionally movies get remade decades later, but most do not. Likewise, popular books like *The*

Hunger Games are not rewritten by other authors. Nor can video game makers freely copy each other's work. But you may have noticed that there are often many versions of a popular song. That's the "mechanical license," going back to the day of player pianos. Once a song has been recorded and published, anyone may make a version—*provided* that he pays a certain royalty to the copyright owner. No permission needed, so this is a compulsory license. In practice, most get a license through the Harry Fox Agency, which represents most U.S. music publishers. Getting a license through Harry Fox is simpler than following that statutory procedure. Like the first sale doctrine, this is a limited right. It allows only recording and selling a version. To make an adaptation, or to perform the song in public, or to use it in a movie soundtrack requires permission from the copyright owner.

A music recording often involves two copyrights: the copyright in the musical work (goes to the songwriter unless otherwise agreed—and often music companies get the copyright by contract), and the copyright in the sound recording (goes to the author of the sound recording, which could be the musician or the producer, and, again, is often assigned by contract). So someone wishing to use the recording may often need a license from two copyright owners.

Questions

1. The makers of the movie *Iron Man* want to use "Iron Man," the Black Sabbath heavy metal song, over the closing credits. If the song's copyright owners are reluctant, can the moviemakers simply use the compulsory license, pay the fee, and use the song?
2. Can Beethoven's works be under copyright?

Answers

1. The compulsory license only authorizes someone to make her own recording and distribute it. It does not authorize using the Black Sabbath recording, nor does it authorize using the song in a movie. The song does appear in the movie, so it would appear the parties found common ground.
2. Beethoven's works are far too old to be under contract. But sound recordings of the works may be under copyright.

Open source licensing, a.k.a. free software, and Creative Commons licenses[6]

> A great high wall there tried to stop me
> A great big sign there said private property
> But on the other side it didn't say nothing
> That side was made for you and me.
>
> Woody Guthrie, "This Land Is Your Land" (1945)
> (copyright not renewed, so in public domain)

People give intellectual property away. Ada may write some code and distribute it almost copyright free, under a free software or open source license. The best known is the GPL, the GNU General Public License.[7] Ada's license would allow anyone to make copies, use the software, adapt the software, and distribute adaptations. Ada does not abandon her copyright but rather distributes copies subject to permissive license terms. She may require that others not put restrictions on the copies of the software. Some free licenses are more permissive than others. The Python SFL and MIT licenses say, in effect, here's some software, do whatever you want with it. Ada may also require that anyone who adapts or redistributes the software must give her credit and avoid attributing modifications to her, which protects her reputation.

Free licensing has grown to other types of works. Open source hardware, such as the Arduino, allows free adaptation and innovation, such as meetups at the London Hackspace. Members of the London Hackspace share their code and hardware design, both face-to-face and online. Music, literature, and all the other creative arts flourished centuries before copyright, so it is no surprise that many artists do not rely on copyright. Reggae music, some say, developed in Jamaica as "open source" music.[8]

The best known free-ish license for distributing books, music, and the like is the Creative Commons license. Creative Commons made it quite easy for artists to create copyright licenses. The CC license tool allows the artist to tailor the permission she gives. The share-alike option requires users to put their works under a similar license, as the GPL does for software. The artist can also choose whether to allow commercial uses of his work, whether to allow others to modify his work, and whether to require others to give him attribution when they use his work. After thousands of artists had used the tool, Creative Commons dropped the no-attribution option.[9] People never choose to allow their work to be used without attribution. Authors will sometimes cede their rights to disseminate their work and they may allow others to use their work and even modify it. Indeed, authors may allow others to make money off their work. But none will surrender the right to get credit for what they have created—especially today, when reputation is a key economic factor.

The CC licenses, the GPL, and other commons licenses put intellectual property in a new light. Inventors and authors can use their intellectual property to keep their works effectively in the public domain. The parties controlling CC and GNU also guard their own rights of attribution. The CC license, for example, cannot be made revocable. The CC license creation tool drafts an irrevocable license, without the option for the author to authorize use of her work but with the right to withdraw permission. Unlike the no-attribution option, a termination right is attractive to many authors. The reason it is not offered is to protect the reputation of CC licenses. If even some CC licenses were terminable, then other creators and distributors would be less likely to rely on CC-licensed works. The GNU license likewise guards against variation. It provides:

Copyright © 2007 Free Software Foundation, Inc. <http://fsf.org/>
Everyone is permitted to copy and distribute verbatim copies of this license document, but changing it is not allowed.

Just as manufacturers rely on trademarks and patents to craft a market presence for their product, so free licensing organizations control their creation.

There is indeed considerable competition among free licenses. Someone ready to give his work away could use the GPL, a CC license, the Artistic License, the MIT License, or many others—or draft his own license. In some areas, freely shared works may replace proprietary works. Intellectual property law, ironically, is proving key to encouraging the sharing of works free of intellectual property.

Question

Some Creative Commons licenses on photos, music, and writings provide that anyone can copy, distribute, or adapt the work without a fee—but not if he makes a commercial use of it. Is that a free license?

Answer

Opinions differ. Some see a "no commercial use" limit as preventing free-riding on free culture. Others think that limits on use mean the work is not free.[10] What do you think?

Ideas Are Not Copyrighted

Copyright protects creative expression, not ideas or functional matter.

Copying ideas does not infringe copyright

Harry Potter and the Philosopher's Stone is under copyright for at least another 70 years. Without J. K. Rowling's permission, it would infringe copyright to sell a book about an orphan named Harry, unknowingly a wizard living with distressing Muggles, who is admitted via owls to Hogwarts, boards a train at Platform 9¾, joins Gryffindor House, learns wizardry and Quidditch, and battles Voldemort. But copyright does not apply to ideas. Anyone may freely copy ideas from the *Harry Potter* book. Anyone could write her own book about a young boy who learns he is a wizard and attends a school to learn magic. Rowling's ideas are not protected by copyright, although her expression of those ideas is. Many magic-themed books followed *Harry Potter*.

Imagine that an author wrote a short story about an authoritarian government distracting the masses with gladiatorial contests among children. Imagine (contrary to fact) that Suzanne Collins read the story and used that idea as the basis for her blockbuster *The Hunger Games*—expressing the idea with completely different characters, events, and plotting. Collins would not infringe the copyright in the story. Copying, even stealing, ideas does not infringe copyright, as many unhappy authors have learned. The author of *Lokey from Maldemar* alleged to no avail that Steven Spielberg purloined her idea of an alien stranded on Earth in making *E.T.: The Extra-Terrestrial*.[1] When someone alleges simply that his idea was copied and made into a movie/book/play/musical, the courts need not even look into whether the alleged copying occurred. Even if the idea was copied, there would be no copyright infringement.

The publishers of *Superman* could not prevent others from copying the idea of a superhero with a secret identity. If one animated penguin movie is successful, others will surely follow. The 1970s saw a succession of disaster movies with literal titles: *Earthquake* (1974), *Meteor* (1979), *St. Helens* (1981), *Hurricane* (1979), *The*

Towering Inferno (1974), *The Black Hole* (1979). Not long after came a trend of parody movies, many using the easy target of disaster movies. Ideas for and from books are also readily imitated.

Even more specific ideas are not protected by copyright. The idea that makes a joke funny, the idea that animates a poem, the idea that distinguishes a professor's journal article: all may be copied without infringing copyright. It may be plagiarism, or bad form, or a breach of ethics, but it is not copyright infringement. Comedians, poets, and professors persevere, nevertheless, showing that forces other than copyright motivate creativity. Likewise, forces other than copyright may deter copying: comedians that steal jokes suffer scorn and repercussions from their peers, as may pilfering poets and professors. As someone observed, "It's tribal. If you get a rep as a thief, it can hurt your career."[2]

Copyright prohibits copying books, songs, software, and artworks, with the goal of producing books, songs, software, and artworks. Copyright would exact too high a price if it prohibited copying ideas. Copyright balances the incentives and control given to authors by allowing free copying of ideas. The theory is that an idea can be expressed in many different ways, so allowing one author a copyright leaves plenty of room for others to work the same land. Otherwise, the first author of a detective novel, romance, or science fiction story would have control over the entire genre. The first academic to publish a new theory in science, a new mathematical result, or a new insight in economics would have the right to that idea for her lifetime plus 70 more years.

Jason FoxTrot could not copyright the numbers one and zero, and so collect from anyone using computers. *Dawn of the Dead*, the movie, was not infringed by *Dead Rising*, the video game copying the idea of zombies hungrily roaming a shopping mall, battled by humans with whatever consumer items, from guns to lawnmowers, were available.[3] "Pi Symphony," which assigned a musical note to each digit from zero to nine in pi, was not infringed by "What Pi Sounds Like," which used the same approach but with different rhythms and harmonies. An acute professor quickly noted that a copyright in pi would be "irrational."[4] Easy palindrome: no IP pi on.

Imitation as a Super Form of Flattery

To decide if the copied material is protected expression or not, the key is whether the copier could use the idea without copying so closely. A super example: The first *Superman* comic in 1938 was quickly copied by *Wonderman*. The ideas in *Superman* could be freely copied, but *Wonderman* copied too closely in the judge's view:

> The attributes and antics of "Superman" and "Wonderman" are closely similar. Each at times conceals his strength beneath ordinary clothing but after removing his cloak stands revealed in full panoply in a skintight acrobatic costume. The only real difference between them is that "Superman" wears a blue uniform and "Wonderman" a red one. Each is termed the champion of the oppressed. Each is shown running toward a full moon "off into the night," and each is shown crushing a gun in his powerful hands. "Superman" is pictured as stopping a bullet with his person and "Wonderman" as arresting and throwing back shells. Each is depicted as shot at by three men, yet as wholly impervious to the missiles that strike him. "Superman" is shown as leaping over a twenty story building, and "Wonderman" as leaping from building to building. "Superman" and "Wonderman" are each endowed with sufficient strength to rip open a steel door. Each is described as being the strongest man in the world and each as battling against "evil and injustice."[5]

Wonderman was dispatched. But the ideas in *Superman* have been mined in the following decades. Many superheroes have appeared, with special powers, secret identities, and vulnerabilities. If they do not track *Superman* too closely, they do not infringe, just as, in fact, Superman was able to build on ideas before him. Superman was not the first to have an Achilles' heel. Nor was Superman infringed by a comic version in the show *The Greatest American Hero*:

> [Ralph Hinkley] had Superman-like abilities in a decidedly unSuperman-like way. For example, Hinkley has the ability to fly, but does so carrying a large lantern and, on occasion, waves his arms wildly to maintain his course or crash lands into a treetop. He has superhuman strength, but, before succeeding in crashing through a wall to surprise the villains, his first attempt nearly knocks him out. Though protected from bullets by his suit, he cringes and covers his face with his arms when shots are fired.[6]

Superman looks and acts like a brave, proud hero who has dedicated his life to combating the forces of evil. Hinkley looks and acts like a timid, reluctant hero who accepts his missions grudgingly and prefers to get on with his normal life. Superman performs his superhuman feats with skill, verve, and dash, clearly the master of his own destiny. Hinkley is perplexed by the superhuman powers his costume confers and uses them in a bumbling, comical fashion. In the genre of superheroes, Hinkley follows Superman as, in the genre of detectives, Inspector Clouseau follows Sherlock Holmes.

1. Could you closely copy the following without infringing copyright?

   "Since the author of the piece has only existed for 0.00000000028571% of the age of the universe, it is a statistical certainty that he wrote it before he was born!"[7]

2. The author of the 1997 work *The Adventures of Willy the Wizard—No. 1 Livid Land* alleges copyright infringement by J. K. Rowling's *Harry Potter and the Goblet of Fire*:

   "[B]oth protagonists are famous male wizards, initiated late into wizarding (in pre/early adolescence), who receive formal education in wizardry, and are chosen to compete in year-long wizard competitions."[8]

   If (a big "if") Rowling copied, did she infringe?

3. After disaster movies like *The Towering Inferno* (skyscraper), *The Poseidon Adventure* (cruise ship), and *Earthquake* (Earth), Hollywood sought new disasters to exploit, notwithstanding the end of the civilized world in *Where Have All the People Gone?* Someone thought, "Zeppelin!" *The Hindenburg* copies liberally from the book *Who Destroyed the Hindenburg?* The film cribbed historic facts from the book author's years of research and his inventive theories about who and what caused the zeppelin disaster. Could there be a new movie idea here, *Copyright Infringement*, perhaps with a jury of zombies?

Facts and Ideas in a Book Are Not Copyrighted

4. *Abie's Irish Rose* jerks tears with an updated Romeo and Juliet story, a secret marriage between a Jew and a Catholic. *The Cohens and Kellys* imitates the interfaith strife, but in a comic

vein, with twin babies confusing grandparents. Copyright infringement?

Plot Ideas Are Not Copyrighted

5. Is the artwork for *Moscow on the Hudson* infringement on *The New Yorker* cover?

6. Homage or infringement?

Copying Too Closely?

7. Again, homage or infringement?

"Wing Tips Over the Edge" Copied by Advertisement

8. "Good artists borrow, great artists steal." Quoted from Picasso (or maybe Oscar Wilde or T. S. Eliot . . .).[9]

Ma Jolie by Picasso   *Man with Guitar* by Braque

Answers

1. Yes, although it would be good form to attribute the source. One could write the sentence in slightly different words and numbers, but such pointless variations would detract from the excellent idea.
2. As the judge put it, the alleged copying involved only "a general sketch of a character (i.e., an unprotectible idea)."
3. Only ideas (theories) and noncreative material (facts) were copied, so no infringement exists. Ironically, copyright gives less protection to perhaps more valuable works requiring research and theorizing exactly because such material should not be owned.
4. No infringement, just copying of the idea.

5. Defendants argued that they simply stole the idea of a New York–centric view of the world. The court (in New York) held it to be copying of the creative expression of the idea.
6. Well, both. Copying not just the idea of a corner of a jazz bar, but the specific creative implementation, such as lighting and composition, using exactly the same corner.[10]
7. Copying of the *idea* of a businessman considering a leap, along with elements necessary to implement the idea. No infringement.
8. Cubism is not copyrighted.

Contracts can protect ideas—a little

Copyright does not protect ideas. If someone with a valuable idea seeks to protect it, the law offers only a thin reed, a contract. If a scriptwriter thinks she has a fantastic idea for a movie, she can offer to discuss it with a movie studio subject to a nondisclosure agreement. Likewise, an author can agree to show his novel to potential publishers only on the condition that they agree not to use his ideas in any other works they publish. That contract will give some protection.

But studios and publishers and other purchasers of ideas often won't sign nondisclosure agreements. Suppose a screenwriter offers to discuss an idea only if the studio agrees to pay if it uses the idea. The studio may think, "Her idea might be one we are already working on, or would get from someone else. Or it may be so vague that lots of things we are working on could be similar. The chance to hear the idea is not worth the risks." Studios and publishers often have a policy to return unsolicited scripts and manuscripts unopened.

Even if a nondisclosure agreement is signed, it is binding only on the parties to the agreement. Others are free to use the idea. In order to commercialize it, at some point the idea will be disclosed. One cannot keep the ideas in a book or movie confidential after publication.

Not to say that one with an idea should despair. Businesses built on unprotected ideas succeed all the time. The person with the idea may be first to market or execute the idea better. Keeping an idea secret may prevent copying, but it also prevents discussing it with others and improving it.

Functional matter cannot be copyrighted

Houdini performs an ingenious illusion, appearing to cut his loyal assistant in two. If another magician copied the trick, there would be no copyright infringement. Instagram develops methods to seamlessly transmit photos. A start-up could copy the methods without infringing copyright. The rules of card games, the engineering design of buildings, systems for analyzing data, designs of tools: such functional matter is not protected by copyright. Copyright protects creative expression even though the creativity may go into inventing functional matter.

Magic tricks, methods of transmitting photos, card games, engineering designs, systems to analyze data, and tool designs all fall outside copyright protection. Houdini can patent his trick if he can get through the stringent patent application process and be content with only 17 years of rights. Copyright cannot be used to get patent-like protection on useful things and processes.

A sculptor created wire sculptures from, as a judge later put it, a "continuous undulating piece of wire."[11] A cyclist friend stated that they would make great bike racks. The sculptor, being practical, adapted the sculpture's design into a sinuous bike rack and sold it under the Brandir name. When other companies copied the ingenious design, Brandir sued—but lost. Copyright does not protect functional things like bike rack designs. The sculptor's bike racks had creative, aesthetic aspects, but they were not separable from the functional elements. A change in the creative design (in the size of the pipe or the width of the curves) would also affect functionality—how the bikes fit in the rack.

Functional items, like furniture, clothing, toys, tools, cell phones, puzzles, and beyond are not copyrightable unless the creative expression is somehow separate from the functional elements. A doll is a sculpture, but it is functional. The Hula Hoop, the Rubik's cube, and the Frisbee are all functional, and so were not protected by copyright. To protect functional matter, a party must seek a patent—either a utility patent on an invention (discussed shortly) or a design patent on the ornamental aspect of a useful article (also discussed shortly). Rubik patented his cube but only in Hungary, which limited his protection in the United States. The *trademark* Rubik's Cube®, however, was enough to build quite a market. The options of copyright, patent, trademark, and trade secret law often offer more than one bite at the apple.

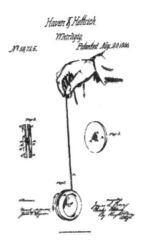

The Yo-Yo Is Functional, Therefore Cannot Be Copyrighted—But It Was Patented

Many toys are not copyrightable because they are functional. For the same reason, many toys have been patented. The best toy, however, is an unpatentable product of nature: the stick, as noted by *Wired*: "pro: grows on trees; con: can poke eye out."[12]

PATENTED MAY 9 1972

3.660.926

SHEET 1 OF 2

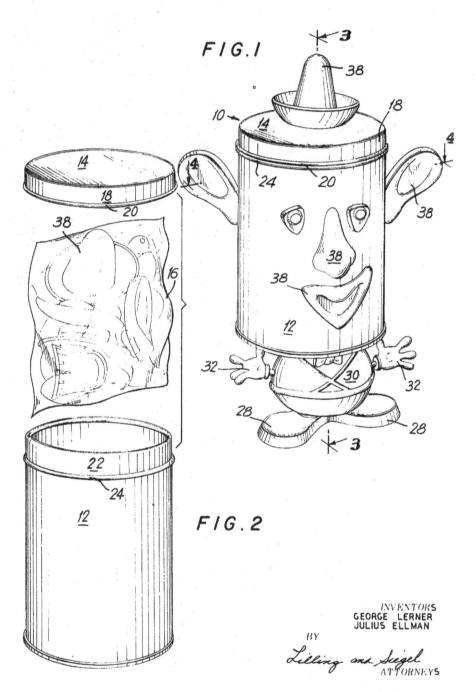

FIG.1

FIG.2

INVENTORS
GEORGE LERNER
JULIUS ELLMAN
BY
Lilling and Siegel
ATTORNEYS

Mr. Potato Head—Functional, but Some Would Say
He Has Separable Creative Elements

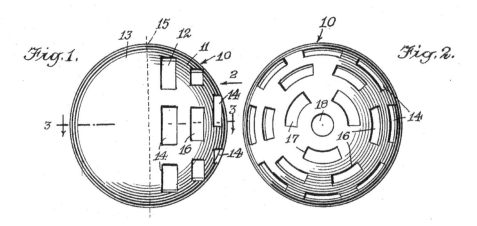

Jan. 1, 1957   W. F. BLAMEY, JR., ET AL   2,776,139

GAME BALL

Filed Feb. 18, 1954

The Wiffle Ball Cannot Be Copyrighted Because It Is Functional (Changing Any Element Would Change How It Works because of Gravity and Aerodynamics)

Questions

1. Can a lamp base with a ballerina statuette be copyrighted?
2. Can the terms of use for a website be copyrighted?

Answers

1. Yes. The statuette aspect is separable from the functional role of the lamp base.
2. No. The language of a contract would have little, if any, creative expression. Rather, it is likely to be made up of clauses that are functional, are ideas, or reuse language from other contracts.

Is software protected by copyright?[13]

Google wanted its Android operating system for phones to run apps written in Java script. Google negotiated unsuccessfully for a license with Oracle, the owner of Java. Instead of copying the code for the software that runs Java, Google wrote its own Java virtual machine and copied only the APIs (application programming interfaces). In English, Google just copied the short headers, like "java.lang.Math.max," that Java uses to interact with other programs. Sued by Oracle, Google argued that the APIs were functional matter, not protected by copyright. The trial court ruled in Google's favor, but the appellate court reversed and sent the case back for further consideration. That example shows that the application of copyright to software is unsettled.

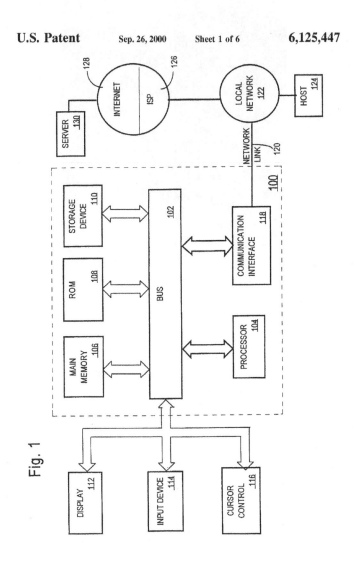

Oracle Also Brought Java Patent Claims against Google

Software may be protected by copyright. Program code is a literary work, a graphical user interface may be a pictorial work, a sequence of scenes could be an audiovisual work. But software is quite functional: it does things. Nevertheless, copying computer code line for line would be infringement. Looking at what a program does in general terms and copying that is not infringement; it is just copying functionality or an idea. Deciding on the cases in between is difficult, like trying to decide just how closely one novel can copy from another without copyright infringement.

1. Could you copyright a launching device for a Superman doll?
2. Is XIO's *Mino* a copyright infringement on *Tetris* in the images shown here?

| *Tetris* | XIO's *Mino* |

Answers

1. No, because it would be functional and therefore not copyrightable, but it could be patentable.

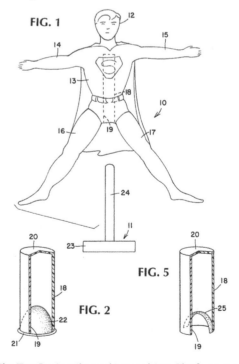

Spector's "Figure-Like Toy Projectile and Launching Platform Assembly"

2. *Mino* copied not just the way that *Tetris* functions but also its creative elements: "[T]he style of the pieces is nearly indistinguishable, both in their look and in the manner they move, rotate, fall, and behave. Similar bright colors are used in each program, the pieces are composed of individually delineated bricks, each brick is given an interior border to suggest texture, and shading and gradation of color are used in substantially similar ways to suggest light is being cast onto the pieces."[14]

By contrast, the early video game *Pong* (patent shown here) was too spare to have creative elements separable from its functional aspects, a basic electronic ping-pong game.

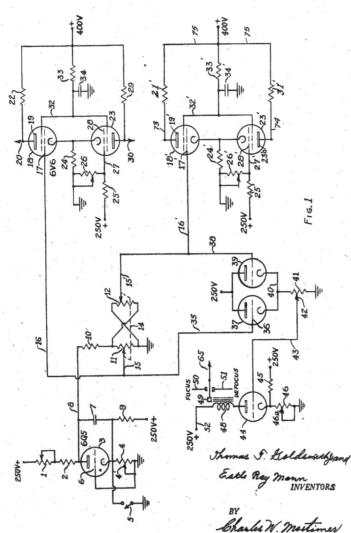

Dec. 14, 1948.   T. T. GOLDSMITH, JR., ET AL   2,455,992

CATHODE-RAY TUBE AMUSEMENT DEVICE

Filed Jan. 25, 1947   2 Sheets—Sheet 1

Thomas T. Goldsmith and
Estle Ray Mann
                    INVENTORS
BY
Charles W. Mortimer

Patent for *Pong*

Design patents

Stepping away from copyright: utility patents are available for inventions of functional things or processes. Another type of patent, a design patent, may be obtained to protect the ornamental design of useful articles—that is, give rights in how it looks. Brandir could have obtained a design patent on the sinuous bicycle rack.

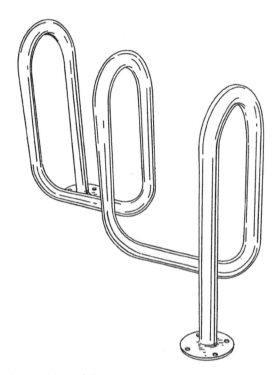

"Bicycle Rack"—But Not Brandir's

Protection is pretty narrow; only close copies infringe. Design patents have played a smaller role than utility patents but are increasing in importance. The aesthetic design of electronic devices is an important market force.

Sony iPhone?[15]

There are more and more design patent lawsuits, such as Apple seeking to prevent others from copying the design of its various iDevices. The complex flow of design influence showed up in the *Apple v. Samsung* suit.[16] Early designs of the iPhone borrowed from Sony products to the extent that Apple even had Sony's logo on some prototypes. In another case, the judge complimented Apple while rejecting Apple's claims, holding that Samsung's alleged iPad knockoffs "do not have the same understated and extreme simplicity which is possessed by the Apple design. . . . They are not as cool."[17]

As with utility patents, there are various mouse design patents, as shown here.[18]

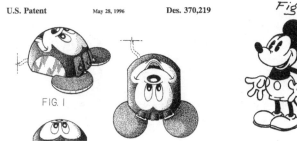

Sylvia Blumer et al., Computer Mouse, U.S. Design Patent 21/656 (1936)

Walt Disney, "Toy or Similar Figure," U.S. Design Patent 82,802 (1930)

Design patents offer protection for works that, like many toys, are functional, so they get limited copyright protection yet may not be sufficiently "new" or "nonobvious" to qualify for utility patents.

Question

Chin-Hung Kwok designs a robotesque toy with springy arms. Creative, but functional, and so no copyright protection. How can the designer discourage copiers?

With a design patent. Here it is:

U.S. Patent   Aug. 29, 2000   Sheet 1 of 6   **Des. 430,237**

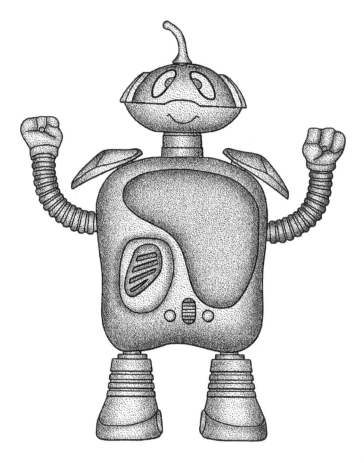

FIG. 1

Chin-Hung Kwok, Toy Figure

5

Formalities: For Want of a ©, the Kingdom Was Lost

Copyright notice: That c in a circle

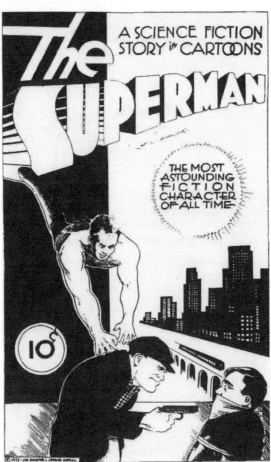

The First *Superman* Comic

Three little words and a ring can mean a lot. Suppose the first *Superman* comics had been published without the copyright notice, "© 1938 Detective Comics." Therein lies kryptonite for copyright. The comics would have forfeited the copyright that they now have until 2033. Anyone could have freely copied the comics, sold copies, and made adaptations (more *Superman* comics, a *Superman* movie or two, *Clark and Lois—The Musical*, *Thus Spake Superman* (opera), *Man and Superman II*).

FIG. 1

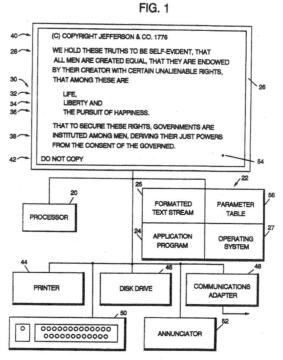

IBM's Patented "System and Method for Managing the Display or the Printing of a Copyright Notice"

Three years are key for copyright: 1909, 1976, and 1989. Here is a mnemonic device for those years: 1909 marked the arrival of the first humans at the North Pole; 1976, the birth of Apple Computer; and 1989, the fall of the Berlin Wall.

Under the 1909 Copyright Act, copyright began when the work was published and lasted for 56 years if the copyright owner filed for renewal at 28 years. But copyright began only if the work was published with a copyright notice: the copyright symbol, the year of first publication, and the copyright owner, such as "© 1929 William Faulkner" on *The Sound and the Fury*. If the work was published without a copyright notice, copyright was forfeit and the work went into the public domain, meaning anyone else could make and sell copies freely. In that case there was no copyright to infringe.

In 1976 Congress revised the Copyright Act (Copyright 4.0, in light of the Copyright Acts of 1790, 1831, and 1909). The 1976 act expanded copyright, to apply not just to published works but also to unpublished works. As of 1978, letters, doodles, diaries, and hobbyist artworks were subject to copyright. But the 1976 act left in place the rule that copyright would be forfeit if a work was published without that magical little notice.

The requirement of a copyright notice and renewal was a trap for the unwary. Amateurs that self-published tended to neglect the task of employing a copyright notice. Publishers in the United States were well aware of the rule, and few of their commercial books lacked that simple notice. But foreign publishers were often unaware of the rule. In almost every other country, copyright applied automatically, notice and renewal or not. The leading international copyright treaty, the Berne Convention, forbade formalities as a condition for copyright. Publication abroad without a notice could forfeit U.S. copyright. Many works lost copyright for failure to file a renewal after 28 years.

Such forfeitures became increasingly problematic for the United States in trade negotiations. In the 19th century, U.S. publishers gladly took advantage of copyright-free foreign works, especially from Britain, a convenient supply of English-language texts. In the 20th century, the United States became increasingly an exporter of copyrighted works (books, movies, music, software, games, television programs, and so forth) and thus more solicitous of copyright. In 1989, in order to join the Berne Convention, the United States dropped the requirement of a copyright notice. If *Harry Potter and the Sorcerer's Stone*, published in the United States in 1998, had lacked a copyright notice, it would not have affected its copyright. An author would not lose copyright by omitting those three little words. For authors, this event was kind of like the 1989 fall of the Berlin Wall.

Harry Potter was nevertheless published with copyright notices: © 1998 J. K. Rowling. Ditto *The Remains of the Day*, © Kazuo Ishiguro 1989. Copyright notices are no longer required, but they continue to appear on books, movies, games, software, and photos. The notice may serve as a warning to potential infringers and an offer to potential licensees, like a sign saying, "Private property. Infringers will be prosecuted. To get permissions or negotiate potential transactions, here's who to contact." Or sometimes a notice is just a friendly way to say, "This is who owns the copyright, should you care."

Before 1989 copyright applied mainly to published works. Now it applies to almost anything: published works, but also anything creative we write, draw, or otherwise make. Do you think that is a good thing?

Renewal of copyright

Chaplin Eating His Shoes in *The Gold Rush*

Charlie Chaplin's film *The Gold Rush* was published in 1925. It could have still been under copyright some 90 years later, but Chaplin did not file the required renewal notice in 1953. Consequently, the film went out of copyright and into the public domain that year.

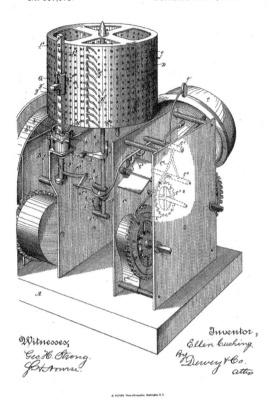

Until 1992 Copyright Was Lost If Not Renewed

Copyright under the 1909 act lasted 56 years, made up of two 28-year terms—provided the copyright owner filed a renewal notice after 28 years. Woody Guthrie's "This Land Is Your Land" (1945) lost copyright in 1973 for failure to renew. Renewal is, admittedly, a low hurdle, but it has been enough to trip up the unwary, careless, apathetic, or confused. Copyright could easily be lost if the author was foreign (hence unaware of the renewal requirement), heedless of legalities (as many authors are), dead, or otherwise occupied (who keeps a to-do list 28 years ahead of time?). In most cases, copyright owners simply did not bother. A 1960 Copyright Office study found that renewal rates were around 7 percent for books, 35 percent for music, 48 percent for maps, and 74 percent for movies.

Renewal continued to be a sinkhole for copyrights until 1992, when Congress made renewal automatic. Works copyrighted after January 1, 1964 (1964 + 28 = 1992) require no renewal to have the full term of 95 years for pre-1978 works. Roald Dahl's 1964 *Charlie and the Chocolate Factory* needed no renewal to be filed in 1993.

Questions

1. *Back in the U.S.A*, © 1959 Chuck Berry: when did he have to file the renewal?
2. *Back in the U.S.S.R.*, © 1968 Lennon & McCartney: when did they have to file the renewal?

3. *The Hunchback of Notre Dame* © 1923 Universal Pictures: when was renewal due?
4. You find a book that was published in the United States in 1961. What are the odds it is still under copyright? (Not that you have to guess; you can search the Copyright Office records online.)

Answers

1. Renewal was due 28 years later, in 1977.
2. For works copyrighted from 1963 on, renewal was automatic.
3. 1923 + 28 = 1951. Universal Pictures neglected to file, so *Hunchback* was in the public domain as of 1951.
4. The book could be under copyright until 2056, a 95-year term, from 28 + 28 + 19 + 20. But if the book was not renewed in 1989 at the end of the first term, it went into the public domain. Only 7 percent of book copyrights were renewed, so the odds are 7 percent. But, of course, failure to renew was not random. Rather, it depended on how much the copyright mattered to the author. The following are the top six sellers from 1961. You can bet their copyrights were all renewed by 1989:

   The Agony and the Ecstasy, Irving Stone
   Franny and Zooey, J. D. Salinger
   To Kill a Mockingbird, Harper Lee
   Mila 18, Leon Uris
   The Carpetbaggers, Harold Robbins
   Tropic of Cancer, Henry Miller

Restoration of foreign copyrights

Picasso's *Guernica* (1937) Restored in 2012 (Coincidentally, the Copyright and the Painting Itself Were Restored That Year)

Picasso painted *Guernica* in 1937, thinking of a recent massacre during the Spanish Civil War. Putting "© 1937 Pablo Picasso" to satisfy the requirements of the copyright law of the United States, a continent away, was not on his mind. He published the work without that copyright notice by freely displaying versions in Europe and the United States. He forfeited his copyright under U.S. law—until 1994.

Foreign authors often lost their U.S. copyright up until 1992. Like Picasso, they could publish without notice (this has been applied even to publication outside the United States). They or their heirs could fail to file a renewal, and their country and the United States might not have agreed to recognize the copyrights of each other. The United States and the Union of Soviet Socialist Republics could barely avoid nuking each other, let alone establish reciprocal copyright relations.

Igor Stravinsky, when living in the USSR, had no U.S. copyright in his compositions. He had a USSR copyright, albeit limited (according to the *Socialist* newspaper in the USSR). After Stravinsky moved to the land of the free, he had copyright in his works, such as his modernist arrangement of his new home's national anthem, "The Star-Spangled Banner." The Boston police, preferring the traditional arrangement, threatened to arrest Stravinsky, the "land of the free" words in the anthem notwithstanding.

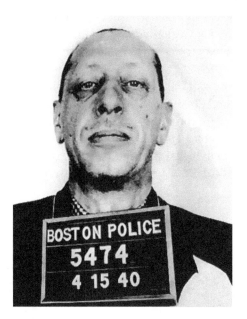

Igor Stravinsky's Photo for Visa to United States (Not a Mug Shot)

As the 20th century progressed, the United States increasingly joined international agreements affecting copyright, patent, and the like, just as it agreed to stop pointing missiles at the USSR. In 1989 the United States dropped the requirement of © in order to join the Berne Convention. In 1994, as part of international trade agreements, the United States agreed to restore copyrights of foreign authors lost due to failure to comply with U.S. formality requirements (© or renewal) or due to lack of copyright relations (such as in the case of the USSR). Thousands of foreign works left the public domain in the United States and are now under copyright as a result. Picasso's *Guernica* and Stravinsky's compositions were among them. The restored copyright receives the remainder of the copyright term it would have had. Foreign works published before 1923, like Stravinsky's "The Rite of Spring" (1913) or Picasso's portrait of Stravinsky (1910), shown here, are not under U.S. copyright.

Portrait of Stravinsky by Picasso (1910)

The restoration of copyrights for foreign authors hit hard some who had relied on them, such as music publishers, educators, and providers of foreign films. Some argued it restricted speech without providing any incentive to authors. The Picasso of 1937, however, cannot now be persuaded to work any harder. But the Supreme Court again upheld Congress's latitude in changing copyright law.

Questions

1. *The Third Man*, a British film about Soviet and U.S. spies in Berlin, lost its U.S. copyright in 1977 due to failure to file a renewal. Case closed?
2. Mick Jagger and Keith Richards of the Rolling Stones, part of the "British Invasion," published "(I Can't Get No) Satisfaction" in 1965. When, if ever, did they have to file a renewal? Suppose they forgot—would there be any satisfaction to be had?

Answers

1. *The Third Man* lost copyright in 1977 for failure to renew, but it regained copyright in 1994 as a foreign work that lost copyright due to compliance with U.S. law's formalities.
2. Works copyrighted after January 1, 1964 (1964 + 28 = 1992) require no renewal.

Registration of copyrights

U.S. Copyright Office Website

An engineering student downloaded and shared some 30 songs. Because the copyrights were registered, the jury could award some $750 to $150,000 per work infringed. The jury settled on a figure somewhere in the middle, $22,000 per song for a total of $675,000. The verdict was affirmed on appeal. The engineer may also have to pay the attorney's fees incurred to sue him. Had the engineer infringed 30 nonregistered copyrights, the award would have been only actual money lost: $30, total.

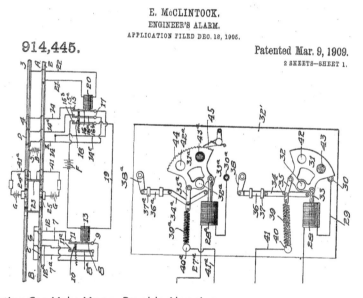

Registration Can Make Money Payable Alarming

Copyright attaches as soon as a work is put into tangible form. Registration is not required. But registration makes the copyright stronger to enforce. Infringers may be ordered to pay $750–$150,000 per work infringed (more if actual money lost is shown), depending on how evil the infringer is, plus attorney's fees (it is galling to pay the attorneys that sued you).

Registration also gives some advantages in proving infringement, as well as putting the copyright in the public record, so others know whom to contact if they wish to license the work. Once a work is registered, parties can also record related documents, such as transfers of ownership or loans, using the copyright as collateral. Recording the purchase of a copyright will protect the buyer if the seller decides to "sell" it again. Unlike cars, a seller cannot deliver an intangible right like a copyright.

Registering a copyright in the U.S. Copyright Office is not expensive or complex. Patent applications require paying thousands of dollars in fees and hiring someone with considerable legal and technical expertise, unless the inventor can handle that herself. Copyright can be registered online for $35, after choosing the appropriate form. Expert representation is always helpful, though. Filing the wrong form or the right form wrongly can have consequences, so an author is well-advised to be well-advised. But copyright registration can be DIY.

The U.S. Copyright Office, along with registration and the former requirements of © and renewal, are distinctive features of the U.S. legal system, although some other countries permit registration of copyrights or require registration for certain works, such as films.

Deposit with registration

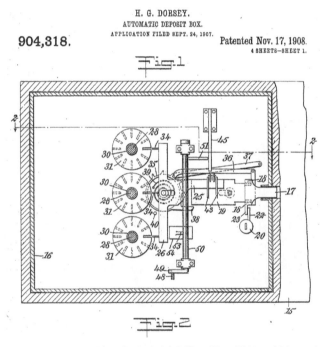

The Copyright Office Would Need More than This Box

Along with the form and fee, registration requires depositing a couple of copies of the work. Authors should even deposit copies of unregistered works that are published, although there is no real penalty for failing to do so ("Stop, or I'll yell stop again"[1]). The copies become property of the United States. The deposit requirement has made the Library of Congress the largest library in the world, a mammoth storehouse of culture. Not everything deposited, however, stays at the Library. Copyright deposit gave the Library copies of works of literature, but also many shampoo bottle labels. The Library chooses which works to retain.

Retention Policy in Action at the Library of Congress (Copyright Deposit
Materials on the Floor in the Thomas Jefferson Building)

Library of Congress Prints and Photographs Division

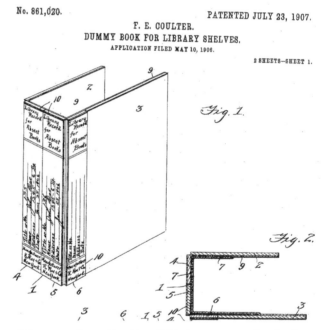

The Library of Congress Does Not Need This (It Has More Books than It Wants)

The deposit requirement has many exceptions. For example, an author
may be reluctant to open to public inspection her diary or the trade-secret
software that runs her weather forecasting service. As another example, a
full-size copy of a giant sculpture would take up a lot of space in the Library
of Congress. The regulations permit redaction of trade secrets, allow certain
sensitive material (such as copies of the SAT test) to be returned after registration, require deposit only of portions of computer programs, and make
allowances for small-scale copies of large-scale works.

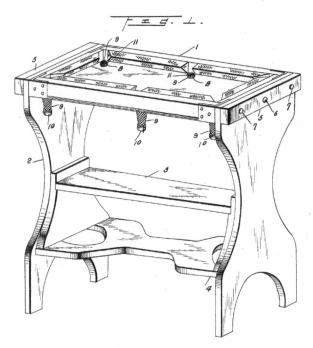

J. D. DAVIS.
COMBINED POOL AND LIBRARY TABLE.
APPLICATION FILED FEB. 1, 1921.

1,410,171.

Patented Mar. 21, 1922.
2 SHEETS—SHEET 1.

The Library of Congress Could Provide Services beyond Books to Read

Much material now may be deposited online in digital form. That material is not readily available to the public. Otherwise, iTunes would go out of business, for people could simply investigate the copyright status of their favorite song or movie over and over again. Likewise, the copyrighted and deposited SAT tests are not available for preview. The Copyright Office has, however, made efforts to make public domain material available. For example, it has a Flickr stream of some fine photography by U.S. employees (noncopyrighted U.S. government works).

6

Fair Use: Excelsior![1]

A teacher makes ten copies of a copyrighted short story to hand out to her class. A book review quotes several lines from a copyrighted novel, and describes key plot points in some detail. A couch potato DVRs favorite copyrighted television shows for viewing at leisure. A teacher Xeroxes a story from the newspaper to discuss current events. A rap group parodies a famous copyrighted country song. The crowd at a baseball game spontaneously sings the copyrighted "Na Na Na Na, Na Na Na Na, Hey Hey Hey Goodbye." Speakers on a beach blanket share Nicki Minaj music with all in earshot. All of these examples potentially fall within copyright's rights (to make copies, distribute copies, adapt the work, publicly perform the work, and publicly display the work). But none infringes copyright; each is fair use.

Fair use balances copyright law. The goal of copyright is to encourage creative expression.[2] If copyright were too strictly enforced, it could work against that goal. Along with the nonprotection of ideas, fair use balances the rights given to authors against the expressive interests of others. Fair use permits uses of works that do not harm the copyright holder's market. It also provides a safety valve in that it can permit uses necessary to allow ideas to circulate freely. Copyright includes fair use and the nonprotection of ideas as "built-in protections for freedom of speech,"[3] on the theory that copyright exists to spur creativity and must give way where necessary to encourage later creative works. Fair use is also a distinctive feature of U.S. copyright law, because few other countries have a general doctrine of fair use (as opposed to specific exemptions for particular uses).

Having said that, we can add that the constraints of copyright serve as an impetus for creativity.[4] Constraints trigger creativity in many ways, from censorship (from movies under the Hays Code to modern writers contending with Web filters, authors can layer meanings to defy restrictions) to literary forms (from the unities in Greek drama to poetic forms such as sonnets, litotes, and haiku—even clerihew) to the internal constraints of a work's plot and characters. In other words, constrained artists sometimes can be more creative. Said poet Philip Larkin, "Deprivation is for me what daffodils were for Wordsworth." Not that we should censor, or should shrink fair use, or put artists in garrets. But this observation suggests looking beyond incentive in copyright policy.

Four-factor test for fair use

To determine whether fair use applies, a court looks at four broad factors: (1) the purpose and character of the use, (2) the nature of the copyrighted work, (3) the amount of copyrighted material used, and (4) the effect on the potential market for the copyrighted work.[5]

The purpose and character of the use

Some favored uses are criticism, comment, news reporting, teaching (including printing multiple copies of a work for classroom use), scholarship, and research. Nonprofit educational purposes are favored; commercial uses are not. Uses that *transform* the work into a different work are more likely to qualify for fair use (but may not—an unauthorized *Harry Potter 8* commercial movie would infringe, as explained in our discussions of the "market harm factor" later in this chapter.

A Machine that Went Up to the Supreme Court

Fair use permitted consumers to use the Sony Betamax video recorder for "time shifting" television programs for later viewing. This was a private, noncommercial use, and there was no evidence that time-shifting harmed the value of the copyrights in television shows.

"Pretty Woman"

Fair use authorized 2 Live Crew's rap parody of Roy Orbison's "Pretty Woman." The parody sharply changed the viewpoint of the song and the nature of the encounter. Such a "transformative use" is likely to qualify as fair.[6]

Questions

1. *Scientific American* reprints a page of a book advocating intelligent design in order to critique its arguments. Fair use?
2. Discount U. scans a textbook and four novels to save its students the price of the books. Educational fair use?
3. Mirror, mirror, on the wall: fairest use of all?

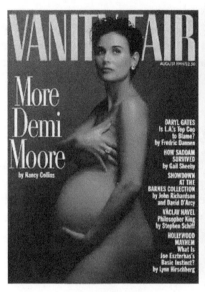 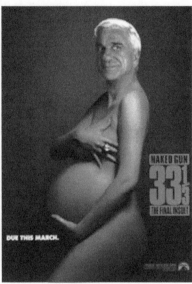

4. A rap song repeatedly used a sample, "Bow wow wow, yippie yo, yippie yea," from George Clinton's iconic "Atomic Dog." The singer did not get a license, although licenses for sampling are quite common in the music business. Fair use?
5. A film documentary promotes "intelligent design" as an alternative to evolution. The film uses 14 seconds of John Lennon's "Imagine" ("nothing to kill or die for, and no religion too") in order to critique the song's worldview. Lennon's widow, Yoko Ono, objects. Fair use?
6. In Carl Hiassen's book *Sick Puppy*, one character drops the lines from rock songs into everyday conversation—but always in a comically wrong way. Fair use?

Answers

1. Fair use. As it would be if *Intelligent Design* magazine quoted a page of *Scientific American* in order to riposte. This would be copying geared to a favored use of criticism and comment (and no market harm—see discussion later in this chapter), and fair use always considers all relevant factors.

2. Not fair use. Educational, but really just a way to lower tuition, a commercial use.
3. The court ruled it fair use—parody commenting on the original. The ensuing litigation further publicized the jest.[7]
4. The court held that fair use did not apply, relying heavily on the standard use of licensing for samples. Other courts might differ, on the theory that using just a snippet and using it in a creative way is a transformative use that is no substitute for the original work.
5. Fair use. Unlike sampling as an element to make another song, this copying was for the purpose of criticism. The use did not build on the creative content but rather was directed at the unprotected ideas. Since it was adversarial, it would likely not have been licensed, so there was no adverse market harm.
6. The utterances are fair use because they take fragments of the quoted works to transform rather than simply quoting them. Nor is there a licensing market for misquotations, unlike sampling in the recording industry.

The nature of the copyrighted work

Works with lots of creative elements (novels, movies, artworks) have a higher level of protection than works composed more of noncopyrightable elements, such as databases or scientific papers (facts and ideas are not copyrighted) or software (functional elements are not copyrighted). For example, it was fair use to make a copy of a computer program in order to reverse-engineer it (to study it, to see how it worked).

Question

> *Art in America* publishes a foldout, high-quality photo of a Jackson Pollock painting, without a license, as part of a special, double-the-usual-price issue on great American art. The copyright holder's negotiations to put the picture on the cover of *Intelligent Art* fall through as a result. Fair use?

Answer

> Not fair use. The work was a highly protected creative work—and the other factors also weigh strongly against fair use (authorial thumb on the scale). The nature of the use was not criticism or education but simply commercial reproduction. By contrast, the *Scientific American* copying mentioned earlier was done in order to accurately portray ideas, which copyright does not protect. Here, the copying was done to exploit the creative expression. And definite market harm resulted from the loss of a potential cover contract with *Intelligent Art*.

The amount of copyrighted material used

The less copying, the more likely it is fair use, especially if the amount copied is appropriate to a favored use, like criticism (quoting in a book review), scholarship (quoting in an academic article), or research (copying a database in order to extract the unprotected facts).

But even a small amount may not be fair use. *The Nation* magazine published prepublication excerpts from former President Gerald Ford's autobiography. The copying was only some 300 words, less than 1 percent of the book. However, fair use did not apply because *The Nation* copied "the heart of the work," the passage in which Ford described pardoning his predecessor, Richard Nixon.

President Nixon with "The King"

President Nixon with Vice President Ford, Secretary of State Henry Kissinger, and Chief of Staff Alexander Haig

Questions

1. The estate of James Joyce threatens to sue for copyright infringement if anyone uses even a few sentences from the author's work. Would it infringe to use several lines literally copied from Joyce's *Finnegan's Wake* in an academic paper, a textbook, or a biography?
2. A book about the Grateful Dead includes low-res copies of copyrighted concert posters. Fair use?

Answers

1. The court held it to be fair use and so not infringing. This is a classic example of how fair use provides breathing room for copyright to serve its ultimate goal, fostering expression.
2. Fair use, the court held. This was commercial, but historical writing, and the limited use was geared to that use, as opposed to big, glossy reproductions.

The effect on the potential market for the copyrighted work

If I write a fan fiction *Harry Potter 8* book and show it only to some friends, no market harm results, and so fair use applies. If I write *Harry Potter 8* and sell it online, that's not fair use.

Questions

1. An art student spends weeks copying a Jackson Pollock painting in order to hone her technique. She exhibits the work at school but does not seek to sell it. Fair use?
2. Are any of these copyright infringement?
   a. A psychology class sings "Happy Birthday to You" to a teaching assistant.
   b. Adelante forwards an e-mail to a friend.
   c. Ada photocopies a news article on encryption and slips it under Turing's office door.
   d. Turisto makes a video of kids running around a museum, and in the background are many copyrighted artworks with new peanut-butter fingerprints.
3. O'Keefe forwards a *New York Times* campaign picture of a politician to a fellow supporter. Fair use?
4. Kritik's scathing review of *Roxygen the Musical* includes a few choice examples of its "insipid lyrics." The review tanks the production. Copyright infringement?

Answers

1. Fair use. The key factor is lack of market harm.
2. All fair use, the key factor being no market harm. Fair use protects many trivial (in legal lingo, *de minimis*) copies in our digital age.
3. Fair use. No market harm, because O'Keefe could have just sent a link (which is the safer thing to do).
4. Fair use. Grievous market harm, but not the sort that counts for fair use. There could be relevant market harm if Kritik wrote a competing musical using the lyrics of *Roxygen*. But the harm from people following Kritik's opinion does not count. Harm from criticism, comment, scholarship, or reporting would not count against fair use, as opposed to providing a market substitute or using a copyrighted work as raw material for another work.

Problems with the four-factor test

The four-factor test could be clearer. There is no point system. Rather, the factors simply guide the court to address the issue: is this particular use a fair use? The following cases, for example, are unsettled:

- A law firm systematically makes copies of scientific articles to keep in its files for reference in filing patent applications.
- A rapper includes seconds-long samples from various songs and improvises on them.
- The Google Books project scans millions of copyrighted books from university libraries, allowing users to search the database of books online, but if the copyright holder objects, users may look only at snippets, not the entire book. So far, courts have ruled for Google Books.

Because of the uncertainty and expense and mental anguish of disputes, people may shy away from making fair use claims. As one scholar put it, that "risk aversion" can, in effect, add to the rights under copyright.[8] Even Google Books has tried to settle, although Google likely thinks the project qualified for fair use.

Questions—Warm-Ups

Fair use can be murky. But the following are all clear-cut cases. Fair use or not?

1. To celebrate Bob Dylan's birthday, a fan makes unauthorized copies of Dylan albums and sells a few hundred on the street.
2. Asked a mathematics question by a teacher, a student simply quotes Dylan: "The answer, my friend, is blowing in the wind."
3. Lumiere shoots video at a family party, catching his Uncle Woody dancing to Bob Dylan music.
4. Scorcese uses Dylan's "Like a Rolling Stone" at length several times, without permission, in the soundtrack of his latest movie, *Like a Rolling Stone*.
5. Walking down Fourth Street, Nikki spontaneously sings Dylan's "Positively Fourth Street."

Answers

No. Yes. Yes. No. Yes.

Question

Abraham makes 50 CDs with copies of 30 of his favorite songs and hands them out at his son Isaac's birthday party. Fair use?

Answer

This would be a noncommercial use, but not just time or space shifting for private use—rather, a substitute for purchased copies. Making and distributing the copies is not fair use. Should it be?

In the unlikely event Abraham is sued, he could have to pay thousands of dollars. See Chapter 7 on litigation.

1. Is rewriting a Dr. Seuss book to tell the story of a murder trial fair use?

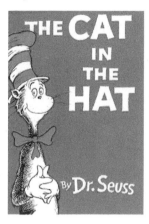 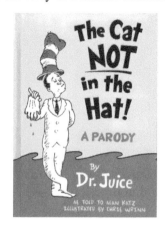

2. Is making a puppy postcard into sculpture fair use?

3. Is borrowing feet fair use?

 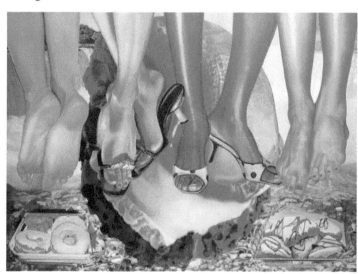

4. Is rewriting photos fair use?

5. Is rewriting *Gone with the Wind* from the slaves' point of view, titled *The Wind Done Gone*, fair use?

Answers

1. Not fair use. Unlike the "Pretty Woman" and *Naked Gun* cases, *The Cat Not in the Hat* did not comment on *The Cat in the Hat* by parodying it. Rather, it used *The Cat in the Hat* simply as raw material to talk about the O. J. Simpson case.
2. A close one. Is this simply using raw material or is it transforming a work—what do you think?
3. This was fair use. Not as a parody, but a transformative use. Part of the image was copied but used quite differently, so this was not simply free-riding.
4. See number 2.

5. This is an effective way to critique the novel *Gone with the Wind*. It is transformative fair use.

Fair use will not permit rewriting a deal. Alfred Hitchcock made the movie *Rear Window* after buying the copyright to a short story, "It Had to be Murder," that appeared in a detective comic book. Hitchcock lost the copyright due to legal technicalities. Fair use did not permit him to continue showing the movie (but a second deal with the copyright holder did).

The Source for Hitchcock's *Rear Window*

Questions

1. The L.A. sheriff's department buys a license to use copies of a software package on 200 computers. It puts the software on 500 computers, but runs a daemon that limits use to 200 computers at any time. Fair use?
2. The United States commissions Frank Gaylord to make a sculpture for the Korean Veterans War Memorial. Frank adamantly does not sign over the copyright. Many visitors take pictures of the sculpture. Fair use? The United States puts the sculpture on a stamp and sells millions of copies. Fair use?

Korean War
Veterans Memorial
2003

USA 37

Answers

1. A deal is a deal. The sheriff's department is making more copies than it agreed to, and this action is infringing. If it wanted different terms, it should have negotiated for them.

2. Visitors taking pictures and posting them online would constitute fair use. It is noncommercial, private use of a public monument. The United States making millions of copies and selling them is not fair use, especially because Frank's deal with the United States left him the copyright.[9]

Questions

Opinions can vary on all of the following.

1. Googlebot crawls the Web, makes copies of all the webpages it can discover, and stores those webpages in a database. If a website does not wish its content to be copied by spiders, it can avoid that by including a robots.txt file with limits. Google makes indexes of all the words on the pages it crawls. When someone searches with Google, Google looks in the indexes for those search terms to find the relevant webpages. Is it copyright infringement when Googlebot copies webpages?

(19) **United States**

(12) **Patent Application Publication** (10) Pub. No.: US 2002/0194161 A1

McNamee et al. (43) Pub. Date: **Dec. 19, 2002**

(54) **DIRECTED WEB CRAWLER WITH MACHINE LEARNING**

(76) Inventors: **J. Paul McNamee**, Ellicott City, MD (US); **James C. Mayfield**, Silver Spring, MD (US); **Martin R. Hall**, Sykesville, MD (US); **Lien T. Duong**, Ellicott City, MD (US); **Christine D. Piatko**, Columbia, MD (US)

Correspondence Address:
Office of Patent Counsel
THE JOHNS HOPKINS UNIVERSITY
Applied Physics Laboratory
11100 Johns Hopkins Road
Laurel, MD 20723-6099 (US)

(21) Appl. No.: **10/121,525**

(22) Filed: **Apr. 12, 2002**

Related U.S. Application Data

(60) Provisional application No. 60/283,271, filed on Apr. 12, 2001.

Publication Classification

(51) Int. Cl.[7] G06F 7/00
(52) U.S. Cl. 707/2

(57) **ABSTRACT**

A web crawler identifies and characterizes an expression of a topic of general interest (such as cryptography) entered and generates an affinity set which comprises a set of related words. This affinity set is related to the expression of a topic of general interest. Using a common search engine, seed documents are found. The seed documents along with the affinity set and other search data will provide training to a classifier to create classifier output for the web crawler to search the web based on multiple criteria, including a content-based rating provided by the trained classifier. The web crawler can perform it's search topic focused, rather than "link" focused. The found relevant content will be ranked and results displayed or saved for a specialty search.

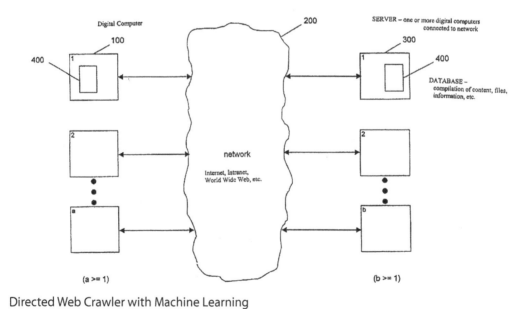

Directed Web Crawler with Machine Learning

2. The National Rifle Association compiles a list of the legislators, along with their contact information, who have particular views on gun control. Gun control advocates manage to get hold of the

database and make copies to use the list in advocating views contrary to the NRA. Fair use?

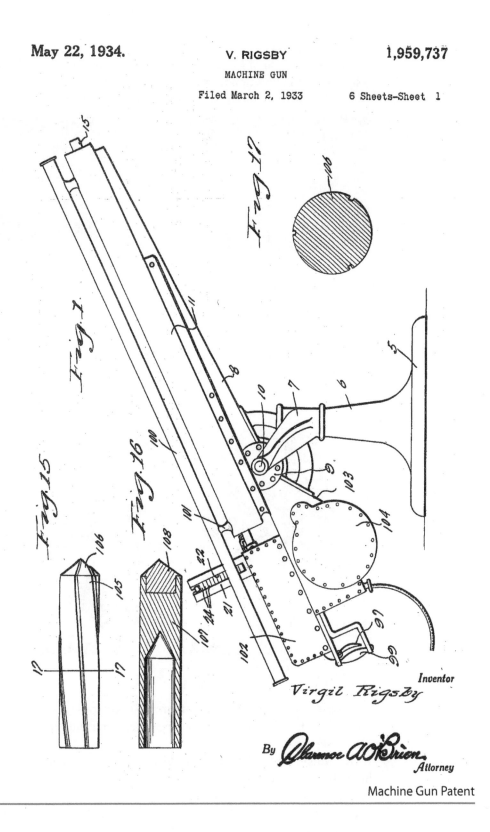

May 22, 1934.   V. RIGSBY   1,959,737
MACHINE GUN
Filed March 2, 1933   6 Sheets—Sheet 1

Inventor
Virgil Rigsby

By *Clarence A O'Brien*
Attorney

Machine Gun Patent

3. A state trooper writes stories in his spare time that reflect his alarming political and social views. He circulates the stories only among a few friends. One of them, however, forwards the stories to a newspaper, which prints excerpts in order to show the trooper's views. Fair use?

4. Abraham Zapruder's home movie camera happened to capture video of the assassination of President Kennedy. An author writing a book setting forth theories of how and why Kennedy was shot included charcoal sketches from stills of Zapruder's movies. Fair use?

Still from Zapruder Film and a Copy Of It

5. An Iowa State student made a 28-minute documentary about a fellow student, a wrestler named Dan Gable. During Gable's run to a gold medal in the next Olympics, the American Broadcasting Company used the student's film extensively, without permission or payment. Fair use?

Pre-TV (*Medieval Wrestling*, Nicolaes Petter, 1674[10])

6. A Google image search shows thumbnail versions of pictures on the Internet. Does Google infringe the copyrights in those pictures?

7. J. D. Salinger wrote *A Catcher in the Rye*, then published very little after that. He wrote a lot of letters, which wound up in various museums. A biography of Salinger quoted extensively from the letters, showing both facts about his life and examples of his style and wit. Fair use?

8. Salinger writes no sequel to *A Catcher in the Rye*, so John California does. Fair use?

9. Lenz shoots a video of her toddler dancing, with Prince's "Let's Go Crazy" playing in the background. She posts it on YouTube. Copyright infringement?

10. How about the New Hampshire resident blaring AC/DC's "Highway to Hell" at the neighbors and lobbing the occasional frying pan?
11. An assassin unwisely writes a book titled *How I Got Away with Murder*. The prosecution makes copies to hand out to the jury. The accused seeks to prevent this taking and distributing of copies of his ill-considered opus. Fair use?

12. A helicopter crew from the Los Angeles News Service (LANS) shot a four-minute video of the beating of a truck driver in the aftermath of the acquittal of police officers accused of assaulting Rodney King. LANS licensed the video to some TV stations. KCAL-TV obtained a copy and broadcast 30 seconds without getting permission. Fair use?

13. Little Publisher quotes several lines from *Finnegan's Wake*. Soon, Little Publisher gets a cease-and-desist letter from the copyright holder's lawyer. Little Publisher seeks support by posting the cease-and-desist letter on its website. Another cease-and-desist letter arrives, contending that posting the letter infringed the copyright in the letter. Does it?

14. Tech Writer, seeking a new job, prints out some examples of her work to show a prospective employer. The examples contain nothing confidential and no trade secrets, but Tech Writer's employer owns the copyrights, as works-made-for-hire. Infringement?

15. Was it fair use for an advertisement for *Moscow on the Hudson* to borrow from a *New Yorker* cover?

Answers

1. Fair use. Google is not using the creative content of the pages or providing market substitutes. Rather, the use is a functional use and does not cut into any market for licensing the creative content. Moreover, Google agrees not to copy pages that have robots.txt refusing copying privileges to its spiders.

2. Fair use. The material has only a thin copyright protection. Names and contact information are facts, and are not protected creative expression. The use is highly favored, noncommercial political activity. There is no market harm to the NRA, or any limitation on the NRA's expressive activity. If someone copied a creatively organized database of the customers of a business and sold copies, that would be different.

3. Fair use. There is no market harm. The stories were unpublished, which weighs against fair use. But the purpose of publishing was to accurately convey the employee's political and social views.

4. Fair use. This is historical writing, along with criticism and commentary (which are favored uses), and the author limited the use of the copyrighted work in a way appropriate to those uses. In particular, charcoal sketches are not a substitute for the stills themselves, but do illustrate the author's theory about the historical event, along with the criticism and commentary.

5. Infringement. ABC preempted the most likely market for the work, use in connection with Gable's Olympic run.[11]

6. Courts have held that fair use applies. The thumbnail versions are not market substitutes. No market harm is apparent. The use is functional, not exploiting the creative content as such but rather helping Internet users find it. Where function, not creative borrowing, is at stake, fair use fosters "copy-reliant technologies."[12]

7. Fair use did not apply. The works were unpublished, although available to museum visitors. The biographer could have copied all the unprotected facts, so literal copying was not necessary. Rather, quoting extensively from the letters allowed the biography also to publish unpublished creative expression, a market that was Salinger's to exploit or not.

8. Infringement. Salinger did not take up that market opportunity, but it was his to control, both financially and artistically.

9. Fair use. Incidental appearances of copyrighted works are inevitable today and are afforded breathing space under the principle of fair use.

10. Fair use. This was noncommercial, with no market harm (unless AC/DC throws frying pans at its fans as the price of admission).

11. Fair use. But nice try.

12. Not fair use. It is a favored use: reporting on a notable incident. But the use was commercial, and LANS did license it to some stations. KCAL-TV cannot unilaterally force a license on LANS.

13. Even assuming the letter qualifies for copyright (i.e., has a little bit of creative menace), posting it would still be fair use. And the next letter, as well.

14. Fair use. Simply showing the work to another would not cut into the market for the work (as opposed to the market for Tech Writer's services, which copyright does not protect).

15. Infringement. This is copying not to criticize or comment on the original but to get raw material.

Copyright Lawsuits

Cease-and-desist letter

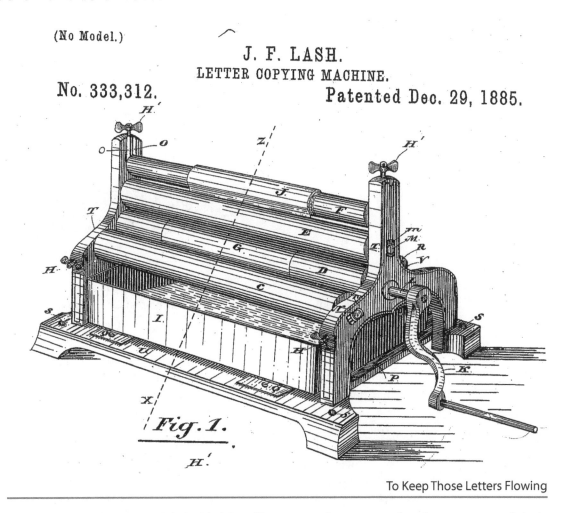

(No Model.)

J. F. LASH.
LETTER COPYING MACHINE.

No. 333,312.

Patented Dec. 29, 1885.

Fig.1.

To Keep Those Letters Flowing

Pepys pastes a *New York Times* article in his blog. He gets good comments, but then a cease-and-desist e-mail comes from the *New York Times*' lawyer. What the heck does that mean?

A letter is just a letter. Anybody can send a letter to anybody. It does not mean that Pepys is infringing. Fair use could apply. It does mean that the *New York Times* objects, claims the use infringes its copyright, and may sue. Sometimes the letter will demand that the activity cease. Sometimes it will demand licensing fees. Some letters are polite, others not so much. Some are form letters; some take days to draft. If Pepys does not comply, the *New York Times* may or may not sue. Sometimes the copyright holder will sue before sending a letter or any other warning.

No copying → no copyright infringement

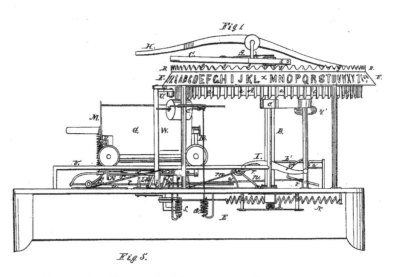

"Improvement in Copying Manuscript": Smoking Gun?

Selle performed his song, "Let It End," a few times in public in Chicago. He registered the copyright and sent a tape and sheet music to several recording and publishing companies, without success. Sometime later, Selle heard a hit song by the Bee Gees (which stands for "Brothers Gibb"), "How Deep Is Your Love," on the radio and in the movie *Saturday Night Fever*. The melody and rhythms of the songs were quite similar. Selle sued the Bee Gees for copyright infringement. At trial, the Bee Gees introduced tapes of their recording session in France, in which one of the brothers created the tune while the other musicians improvised along, gradually making the finished song. There was no copyright infringement: no matter how similar the works, there cannot be copyright infringement without copying. If the Bee Gees did not copy, they did not infringe, even if Selle came up with the tune first (a point often lost on artists of all stripes).

When the defendant denies copying, courts look to three things, as in *Selle v. Gibb*:[1]

1. *Similarity*: The works were quite similar, showing there might have been copying—but it could also be coincidence, like the random similarity between the name of Selle and the brand Sellotape®.

2. *Access*: There was no evidence that the Australian Bee Gees had any access to Selle's song, which had limited distribution in Chicago, before their recording session in France, a fact that undercuts the chances of copying.

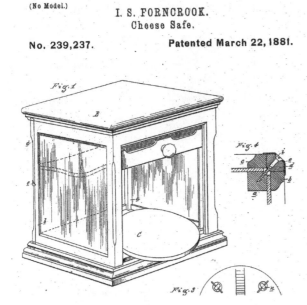

No Access, No Copying

---

3. *Independent creation*: The Bee Gees had a recording of the process of putting the song together without any use of Selle's work, likewise making copying unlikely.

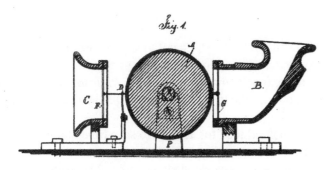

Recording the Creative Process Rebuts Claims of Copying

Disputes about copying often crop up in cases involving successful books, songs, and movies. Many songs have similarities. Story-based works, like books and movies, tend to use similar plots, gags, types of characters. Many an author hears a song, reads a book, or sees a movie and is convinced that it was stolen from his work. Many lawsuits are filed against popular artists. Some are not entirely rational, accusing pop stars of burglarizing the homes of obscure dittyists or using mind-reading technology. Some are well-founded. Several musicians have been found to have, perhaps unknowingly, recycled songs that influenced them in their youth. UCLA's Music Copyright Infringement Resource, at cip.law.ucla.edu, has snippets from dozens of song battles over copyrighted songs (including Selle's song).

Innocent Copying Is Still Infringement

Jean Miélot, also Jehan (d. 1472), Scribe for Philip the Good, Duke of Burgundy from 1449–1467. *Paris, Bibliothèque Nationale. Ms. Fr. 9198, f. 19*[2]

Copying may be innocent or even unconscious.[3] Michael Bolton denied copying his song "Love Is a Wonderful Thing" from the Isley Brothers song that bore the same name a quarter of a century earlier. Bolton convincingly testified he could not remember ever hearing the song. But evidence did show he loved listening to the Isleys in his youth. Lots of access and similarity, together with no evidence of independent creation, led to the Isleys winning a big judgment.

Questions

1. Arthur Dent's Infinite Monkey Drive generates random characters (as from monkeys randomly hitting typewriter keys) until it generates the text of *Harry Potter and the Philosopher's Stone*. Can Arthur now sell the text on the theory that he did not copy from J. K. Rowling?

2. Why is it a good idea for authors to save time-stamped drafts and musicians to save studio tapes?

3. Former Beatle George Harrison releases "My Sweet Lord." Sued by the holder of the copyright in "She's So Fine," George convinces

the court that he had no intention to copy, and rather must have subconsciously copied it from his boyhood memories. Copyright infringement?

Answers

1. It would still be copyright infringement. Arthur copied, albeit via a complicated way. If Arthur truly worked independently and wrote a similar book, that would not be infringement, no matter how similar it was.
2. Saving evidence of the production of a work provides good defense against a possible future claim of copyright infringement—unless the artist actually is infringing, in which case it is like a confession.
3. Harrison sang elsewhere, "No use handing me a writ, while I'm trying to do my bit." But this writ worked. Harrison infringed by copying, even if he did so quite innocently and unconsciously.

Was protected material copied?

Most copyright cases do not have disputes about copying. Usually, there is no doubt that the defendant copied; the question is whether the copying was wrongful. Suppose Jackson made a movie based on a story "The Haggis." Jackson would not be infringing if

- he copied only ideas (nonprotection of ideas).
- he copied only facts or other material not originated by "The Haggis" author, such as material from other works (nonprotection of others' material).
- he had a license from the author, a co-author, or someone who had rights from the author (permission).
- he copied only functional elements (nonprotection of functional matter—more likely in areas like software or fashion).
- "The Haggis" is not copyrighted, or if it was published before 1923, or if it was published before 1989 without the necessary copyright notice, or if it was not properly renewed (pre-1964 works).

Questions

1. Chucky Duckens carefully studies the *Harry Potter* series to see which elements made it successful. Chucky copies many ideas from the books, along with copying many elements that J. K. Rowling borrowed from other books, in writing his own series, *A Tale of Two Copperfields*. Is this copyright infringement? Would it matter if Chucky lied, saying he had never read *Harry Potter*?
2. UCLA's Music Copyright Infringement Resource, at cip.law.ucla .edu, has snippets from dozens of song battles over copyrighted songs (including Selle's song). It has copied from dozens of songs. Did UCLA infringe by the dozen?
3. Suspicious that others are copying its information, *The Trivia Book* includes false nuggets of trivia ("Abraham Lincoln invented Lincoln

Logs"). When the spurious facts show up in other trivia books, *The Trivia Book* sues for infringement. Will it win?

Answers

1. No infringement. Chucky copies only nonprotected ideas and elements copied from others. It would not matter if Chucky falsely denied relying on *Harry Potter*. Failure to give credit is not copyright infringement. It is plagiarism. That may be unethical, but it is not illegal.
2. UCLA did not infringe. Its copying is fair use.
3. *The Trivia Book* will not win. Facts, even trivial ones, are not copyrightable. If they present information as facts, they will not later be able to complain when others copy. Same result if they invented a poem supposedly from the 1800s or a newly found Leonardo da Vinci drawing.

How to sue yourself

Being accused of copyright infringement can cast uncertainty over a project, even if the copyright holder has not filed a lawsuit. To resolve things, the alleged infringer can file a lawsuit for declaratory judgment, seeking a court order that the use at issue does not infringe copyright. An adroit use of this technique involved the works of James Joyce. The Joyce estate liberally threatened copyright infringement actions against anyone who so much as quoted a few words from Joyce's writings, even though such use is the embodiment of fair use. The threats made it difficult for those writing about Joyce to find willing publishers. A literary scholar brought a declaratory judgment action, winning not just a decision upholding the application of fair use, but also an award of attorney's fees, meaning the Joyce estate funded the litigation against itself, thus establishing fair use of its copyrighted works.[4]

Remedies

This domain name associated with the website Megaupload.com has been seized pursuant to an order issued by a U.S. District Court.

A federal grand jury has indicted several individuals and entities allegedly involved in the operation of Megaupload.com and related websites charging them with the following federal crimes:

Conspiracy to Commit Racketeering (18 U.S.C. § 1962(d)), Conspiracy to Commit Copyright Infringement (18 U.S.C. § 371), Conspiracy to Commit Money Laundering (18 U.S.C. § 1956(h)), and Criminal Copyright Infringement (18 U.S.C. §§ 2, 2319; 17 U.S.C. § 506).

One Remedy for Infringement—Shutting Down a Website

The word "remedies" in everyday life summons up thoughts of feeding chicken soup to people with colds or rubbing mud on poison ivy rashes. Legal *remedies* are the reparations ordered by the court. If there is copyright infringement, the defendant may be ordered to pay money and to stop infringing and to take other appropriate action, such as handing over an Internet domain, plus having material confiscated and destroyed. In a criminal case, there may be fines and imprisonment.

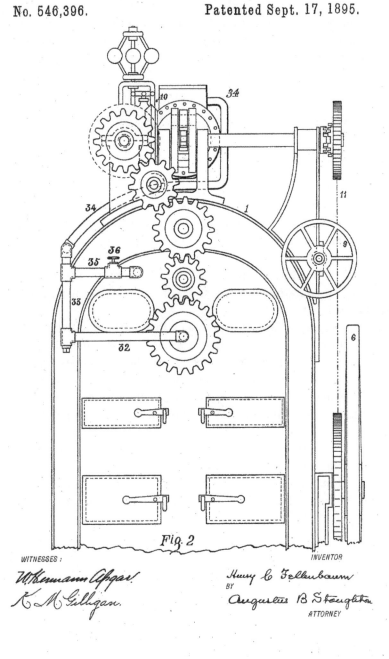

(No Model.)

6 Sheets—Sheet 2.

H. C. FELLENBAUM.
PORTABLE GARBAGE INCINERATOR OR DESTROYER.

No. 546,396.

Patented Sept. 17, 1895.

Fig. 2

WITNESSES :

INVENTOR

Henry C Fellenbaum
BY
Augustus B Stoughton
ATTORNEY

ANDREW B GRAHAM. PHOTO-LITHO. WASHINGTON. D.C.

Bookmobile for Those Infringing Copies

Courts may order remedies to fit the infringement. Where a logo on the Baltimore Ravens' football uniforms infringed, NFL Films was ordered to blur out the logo on old football-highlight films.[5]

Question

Suppose the owner of the Baltimore Ravens had the infringing logo tattooed on his arm. Would a court order it to be excised?

Answer

"The quality of mercy is not strain'd." Reason will temper court orders. As in *The Merchant of Venice*, courts will not order surgery just to remedy legal wrongs.

Show me the money: Twice blest

How much does an infringer pay if the infringer has money? There are two measures: actual damages or statutory damages.

Actual damages: The court may order payment of money lost. The copyright holder may be reimbursed for its lost sales or lost licensing fees, or may get the wrongful profits of the infringer. Suppose an infringing *Harry Potter 8* book is written and sold across the country. The court could order payment of lost sales (if J. K. Rowling had planned a surprise *Harry Potter 8* and the infringement derailed it) or payment of the infringer's profits (hand over the money from the sales, minus the costs). For commercial infringement, such awards can be big.

Statutory damages: Sometimes money lost does not amount to much or is difficult to prove. Suppose copies of a book are distributed without permission, but not for a price. The author cannot show lost sales, and there are no profits to hand over. Copyright law gives another measure if the copyright was registered. The court may order payment in the range of $750 to $150,000 per work infringed. Why require the infringer to pay much more than the money lost? Copyright can be difficult to enforce, so increasing the award may deter infringers (*pour encourager les autres*) and also give copyright holders an incentive to enforce their rights.

The court may also order the losing party to pay the winner's attorney's fees. The possibility of attorney's fee awards encourages a party in the wrong to settle.

Sheriff and Deputies in Court

Library of Congress Flickr stream

1. A grad student downloaded some 22 songs. What would be the money lost? How much could the court award?
2. In which of the following cases would the copyright holder choose an award based on the money lost?
   > Singer's smash hit infringes old-timer's country tune.
   > Obscure blogger copies 50 photos from paparazzi website.
   > University puts dozens of scientific journal articles online without permission.
   > Producer Max Bialystock steals the script for *Funny Boy* (a musical based on *Hamlet*). Bialystock's musical bombs.
3. In the millions of lines of code for the Android operating system, Google copied lots of noncopyrighted functional matter from Sun's Java software. Google copied a few lines of code as follows:

```
private static void rangeCheck(int length, int fromIndex, int toIndex) {
  if (fromIndex > toIndex) {
    throw new IllegalArgumentException(
        "fromIndex(" + fromIndex + ") > toIndex(" + toIndex + ")");
  }
  if (fromIndex < 0) {
    throw new ArrayIndexOutOfBoundsException(fromIndex);
  }
  if (toIndex > length) {
    throw new ArrayIndexOutOfBoundsException(toIndex);
  }
}
```

Millions of copies of Android code have been put in phones and other devices. If that little bit infringed, what would Google have to pay?

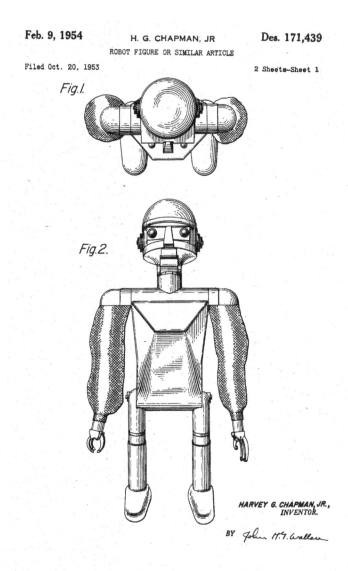

Different Android

Waiting for the Jury

4. In a novel, *Atavistic Avatar*, a robot arrives on Earth in a space-ship disguised as a 1967 Cadillac. The robot looks and behaves as much like a human as its makers could manage, working from radio waves that reached their planet (some decades back, to give time for the programming to reach them, a ship and robot to be built, and to travel here). Despite their advanced technology, the robot's makers could not muster the artificial intelligence for automatic translation of conversational speech. The robot speaks from a vast library of canned phrases gleaned from earthly broad-casts: "Sometimes words have two meanings."; "I'm sorry, Dave. I'm afraid I can't do that."; "Don't worry about a thing, 'cos every little thing is gonna be alright." The character quoted 100 lines from 100 different songs and movies during the course of the novel. Any exposure here? What if it were 100 lines from the same novel?

5. Remember Abraham, who made a party favor CD with 50 of Isaac's favorite songs and gave the CDs to the birthday party guests? Any risk of infringement?

Jury Arriving at Court

Library of Congress Flickr stream

1. Actual losses: about a dollar a song. But the court may award $750 to $150,000 per song. The jury in the case ordered the student to pay $650,000(!), some $22,000 per song. Ouch. So much for the quality of mercy. Appeals are pending.

2. Here are the answers to the question about in which of the following cases the copyright holder would choose an award:

   Singer's smash hit infringes old-timer's country tune: An award in the range of $750 to $150,000 per song would be less than the actual losses if Singer's profits from the smash hit were more than $150,000.

   Obscure blogger copies 50 photos from paparazzi website: An award of $750 to $150,000 per work would be more.

   University puts dozens of scientific journal articles online without permission: Ditto: lots of works infringed and little money lost.

   Producer Max Bialystock steals the script for *Funny Boy* (a musical based on *Hamlet*). Bialystock's musical bombs: Only one work is infringed, so an award not based on actual losses would be $750 to $150,000. But it may be hard to show money lost, because Bialystock made no profits. If, however, we can show the author of *Funny Boy* lost other opportunities as a result, that money would be recoverable.

3. An award of $750 to $150,000 is chicken feed in this case. The code was put on millions of devices, but the award is measured *per work infringed*. Nor was there much in the way of actual money lost for such trivial copying.

4. The novel's author could infringe 100 copyrights, for an award per copyright in the $75,000–$15,000,000 range, or infringe no copyrights, depending on whether a court applied fair use. Scary.

    If it were 100 lines from one novel, then the range is $750–$150,000. Somewhat less scary.

5. If not fair use (more on this later), one award per work infringed is, at a minimum, 50 times $750: $37,500(!). And the high end (very unlikely, but just to show the range) is 50 times $150,000: 7,500,000(!!!). Having said that, no such case has occurred. But it shows what an odd legal situation could result. Much of the U.S. population is potentially on the hook for gigantic copyright damages.

Patentable Inventions: Products and Processes

Whoever invents or discovers any new and useful process, machine, manufacture, or composition of matter, or any new and useful improvement thereof, may obtain a patent therefor, subject to the conditions and requirements of this title.

Section 101, Patent Act

Build a better mousetrap, and the world will beat a path to your door.

Ralph Waldo Emerson (remixed)[1]

What can be patented?

Tammy Edison invents a super-efficient light bulb in 2013. She's entitled to a patent if the invention is new (not previously published or in public use), useful (it works), and not obvious (to a light bulb engineer). Unlike copyright, patent requires the inventor to file an application describing and claiming the invention. If Tammy's patent application is approved, then for about 17 years (depending on how long the approval process takes) she decides who may make, use, sell, or import that super-efficient light bulb.

Patent, like copyright, serves as an incentive for innovation. Thomas Jefferson famously expressed skepticism about property in information: "He who receives an idea from me, receives instruction himself without lessening mine; as he who lights his taper at mine, receives light without darkening me."[2] But Jefferson soon said that the first U.S. patent law had "given a spring to invention beyond his conception."[3]

What can be patented? Products and processes. A *patented product* can be (1) a machine (the telephone), (2) an article of manufacture (a button), or (3) a composition of matter (Claritin, the allergy drug). A *patented process* can be (1) a method of doing something (pasteurizing milk), (2) a method of using something (using a laser to etch on rock), or (3) a method of making something (making nanocircuits). We will look at patented products first.

Questions

1. Could the word "supercalifragilisticexpialidocious" be patented?
2. Could a propagated electronic signal be patented?

Answers

1. A word is not a product (machine, article of manufacture, or composition of matter) or a process, so it is not patentable, as would be a hyperintelligent shade of the color blue.
2. An electronic signal is likewise not a product or process, so it is not patentable, although electronic signals can have many high-tech uses. The process for making the signal could be patentable.

Products

As explained earlier, patentable products can be machines, articles of manufacture, or a composition of matter.

Machines

Wright Brothers (1905)

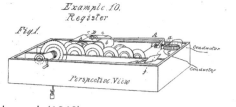

Morse, Telegraph (1840)

T. A. EDISON.
Electric-Lamp.

No. 223,898.   Patented Jan. 27, 1880.

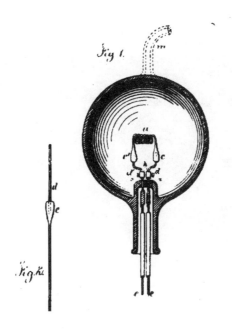

May 17, 1955   E. FERMI ET AL   2,708,656

NEUTRONIC REACTOR

Filed Dec. 19, 1944   27 Sheets—Sheet 1

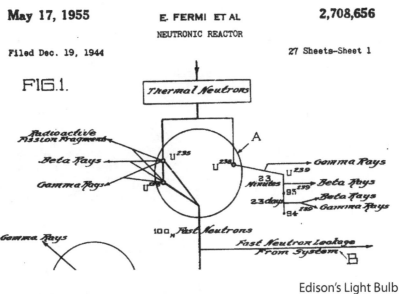

Edison's Light Bulb

To illustrate pioneer patents, the Supreme Court listed famous machines: the "telegraph (Morse, No. 1,647); telephone (Bell, No. 174,465); electric lamp (Edison, No. 223,898); airplane (the Wrights, No. 821,393); neutronic reactor (Fermi & Szilard, No. 2,708,656); laser (Schawlow & Townes, No. 2,929,922)."

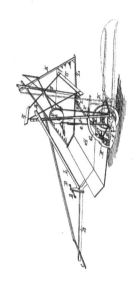

Cyrus McCormick, Improvement in Machines for Reaping Small Grain

McCormick's reaper moved farmers from rocky, hilly New England to the flat, loamy Midwest.

(No Model.)   J. E. MATZELIGER,   7 Sheets—Sheet 1.
LASTING MACHINE.
No. 274,207.   Patented Mar. 20, 1883.

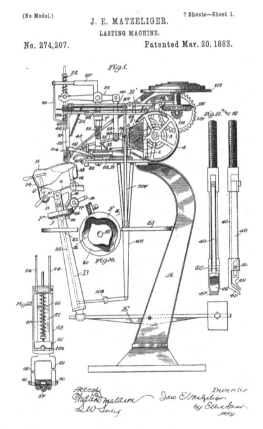

Jan Earnst Matzeliger's Last Spurred Shoe Manufacturing[4]

Oct. 18, 1949.   J. Y. COUSTEAU ET AL   2,485,039
DIVING UNIT

Filed March 10, 1947   3 Sheets—Sheet 2

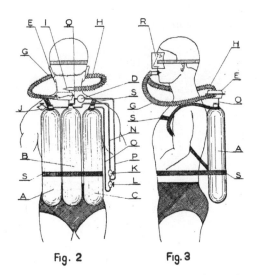

Fig. 2   Fig. 3

Jacques Cousteau and Emile Gagnan, U.S. Patent 3,095,890, Demand Regulator for Breathing Apparatus (1963)

A patentable machine can be any device that makes something, changes something, or does something. Jacques Cousteau's aqualung supplied divers with air at the proper pressure, opening the undersea world to exploration.

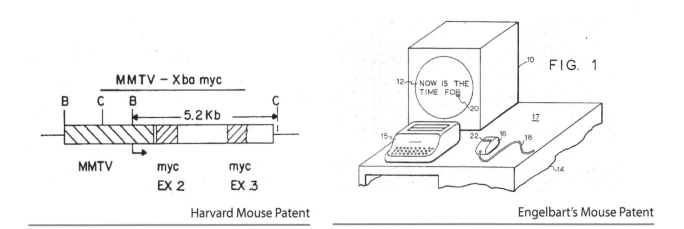

Harvard Mouse Patent   Engelbart's Mouse Patent

A living mouse can be a machine. Engelbart developed the computer mouse in Menlo Park, California, and patented it as "a position indicating control apparatus which is moveable over a surface to provide position control indications." Claude Shannon sought a patent on Theseus, his electromagnetic maze-solving mouse, one of the first electric devices that could change its output based on experience. A living mouse can be patented. One patented machine is the "Harvard mouse," genetically engineered to be susceptible to cancer (hence, also known as "OncoMouse"). Or, as claimed in the patent: "A transgenic non-human mammal all of whose germ cells and somatic cells contain a recombinant activated oncogene sequence introduced into said mammal, or an ancestor of said mammal, at an embryonic stage." Another team patented an NR2B knockout mouse, an especially forgetful mouse with a key memory-related gene blocked.

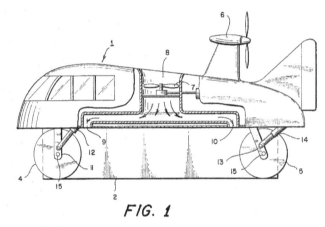

Oct. 4, 1966   C. H. LATIMER-NEEDHAM   3,276,529

GROUND EFFECT VEHICLES

Filed June 11, 1964   4 Sheets—Sheet 1

FIG. 1

"My Hovercraft Is Full of Eels."[5]

Questions

Recognize these machines?

1.

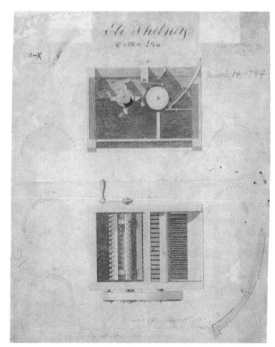

2. No. 808,897. PATENTED JAN. 2, 1906.

W. H. CARRIER.
APPARATUS FOR TREATING AIR.
APPLICATION FILED SEPT. 16, 1904.

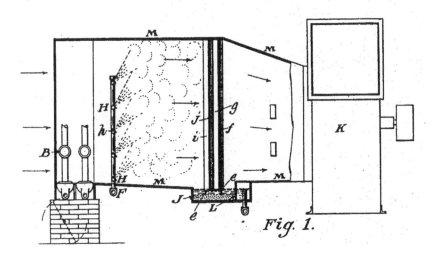

Fig. 1.

3. (No Model.) 7 Sheets—Sheet 2.

W. S. BURROUGHS.
CALCULATING MACHINE.

No. 388,116. Patented Aug. 21, 1888.

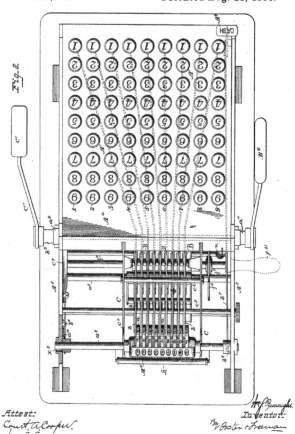

Attest:
Court A. Cooper.

W. S. Burroughs.
Inventor.
By Foster & Freeman

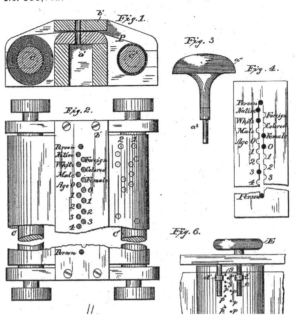

4. (No Model.)

H. HOLLERITH.
ART OF COMPILING STATISTICS.

No. 395,782.

3 Sheets—Sheet 1.

Patented Jan. 8, 1889.

Answers

1. Eli Whitney's cotton gin (*gin* as in engine) shows that technology has good and bad effects: It "not only helped make many people rich on both sides of the Atlantic, but also reinvigorated slavery, turned child labor into a necessity, and paved the way for the American Civil War . . . That is quite a lot of consequence for a simple rotating drum."[6]
2. Air conditioner—another mixed bag.
3. William Seward Burrough's adding machine. William Seward Burroughs II was the Beat Generation author of *Naked Lunch*.
4. A little obscure: Hollerith's punch-card processor, which miraculously processed the data for the 1890 census in only a year, was a forerunner of computers.

Articles of manufacture

More static products are also patentable. Articles of manufacture include all kinds of things made by humans: gloves, shoes, pipes, furniture, helmets, the necktie, Legos, and many, many more.

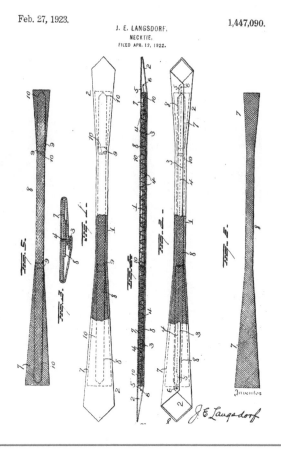

Feb. 27, 1923.

J. E. LANGSDORF.
NECKTIE.
FILED APR. 12, 1922.

1,447,090.

Necktie Patent

Oct. 24, 1961

G. K. CHRISTIANSEN

TOY BUILDING BRICK

3,005,282

Filed July 28, 1958

2 Sheets-Sheet 1

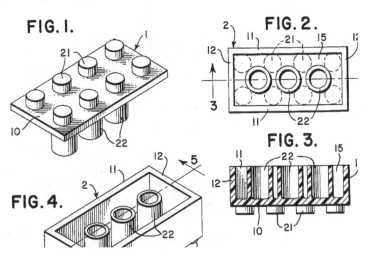

FIG. I.

FIG. 2.

FIG. 3.

FIG. 4.

Lego Patent

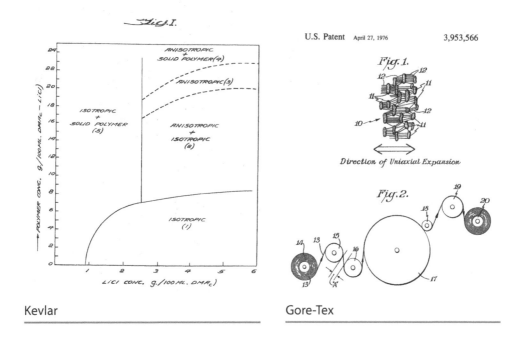

Kevlar   Gore-Tex

Patented articles of manufacture include Kevlar, developed by Stephanie Kwolek, which makes light, bullet-proof clothing, and Gore-Tex, by Robert Gore, which makes mountaineering more high-tech (less of a challenge?) than when Edmund Hillary and Tenzing Norgay climbed Everest in tweed jackets and wool scarves.

BURBERRY

ESTABLISHED 1856

Successor to the Maker of the Coats that Climbed Everest

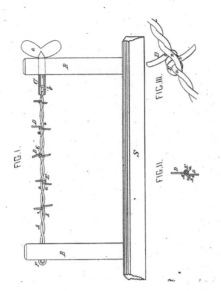

Joseph Glidden's Barbed Wire Changed American Agriculture by Encouraging
Cattle to Remain on the Ranch

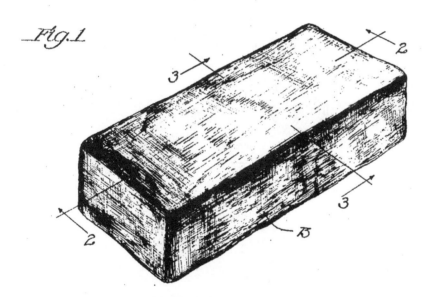

The present invention relates to confectionery and has for its object the production of a commercially practical coated brick
or block of ice cream or the like.

The Ice Cream Bar (U.S. Patent 1,404,539), Like Air Conditioning, Made American
Summers More Tolerable

CIRCUIT ELEMENT UTILIZING SEMICONDUCTIVE MATERIAL

Filed June 26, 1948   3 Sheets—Sheet 1

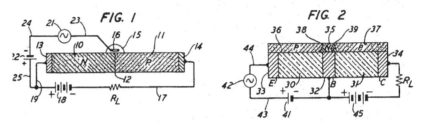

Without the Transistor, the iPhone Would Need Wheels

Questions

Which articles of manufacture have we here?

1.   L. YALE, Jr.
                    POST-OFFICE DRAWER LOCK.

No. 31,278.   Patented Jan. 29, 1861.

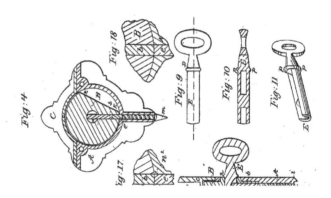

2.   Aug. 12, 1930.   C. BIRDSEYE   1,773,080

ANIMAL FOOD PRODUCT

Filed June 20, 1927

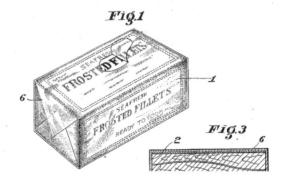

3. No. 748,626.

PATENTED JAN. 5, 1904.

L. J. MAGIE.
GAME BOARD.
APPLICATION FILED MAR. 23, 1903.

NO MODEL.

2 SHEETS—SHEET 1.

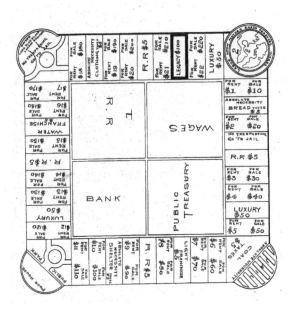

4. Aug. 19 , 1924.

1,505,592

F. W. EPPERSON

FROZEN CONFECTIONERY

Original Filed June 11 , 1924

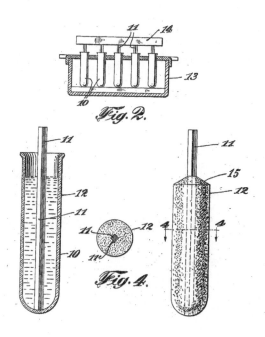

1. Linus Yale's cylinder lock (U.S. Patent 31,278, 1861) is likely the type in any deadbolt or safe you see.
2. Frozen food, the basis of the Birds Eye business empire. The patent claimed the invention in zombiesque terms:

> **2. An article of manufacture, comprising a carton packed with frozen animal tissue which has been frozen therein, said carton being substantially hermetically sealed and having throughout the same undistorted cross-sectional outline as before freezing, the contained animal tissue having in its frozen condition substantially the same cellular structure as it had before being frozen and retaining its pristine qualities and flavors.**

3. Lizzie J. Magie's *The Landlord's Game*, U.S. Patent 748,626, better known as *Monopoly*.
4. Popsicle®.

Composition of matter

A mixture of stuff may be patented, whether material is combined at the molecular level (such as drugs and other chemicals) or on a larger scale (such as the formula for a sports drink).

United States Patent Office

3,056,785
Patented Oct. 2, 1962

1

3,056,785
PURINE DERIVATIVES
George H. Hitchings, Yonkers, and Gertrude B. Elion, Bronxville, N.Y., assignors to Burroughs Wellcome & Co. (U.S.A.) Inc., Tuckahoe, N.Y., a corporation of New York
No Drawing. Filed Mar. 21, 1960, Ser. No. 16,203
8 Claims. (Cl. 260—252)

The present invention relates to a new group of di-heterocyclic sulfides, containing at least one purine moiety. These substances may be represented by the following formula:

2

removing aliquots periodically and examining their ultra violet absorption spectra, since the spectra of the products are quite different from the spectra of the starting materials.

TABLE I

Quantities in gamma per ml.:	Percent inhibition
DDS 4/ml.	15
DDS 8/ml.	25
DDS 16/ml.	55
Compound 10/ml.+DDS 4/ml.	--
Example 1, DDS 4/ml.	64
Example 2, DDS 4/ml.	54
Example 3, DDS 4/ml.	59
Example 4, DDS 4/ml.	81
Example 6, DDS 4/ml.	85
Example 8, DDS 4/ml.	87
Example 9, DDS 4/ml.	69
Example 10, DDS 4/ml.	69
Example 11, DDS 4/ml.	73

U.S. Patent on the Immunosuppressant Imuran

Imuran, invented by Gertrude Belle Elion and George Hitchings, suppresses immune reactions, making it possible to transplant organs. Some patented drugs bring in billions of dollars a year in sales. Such an invention could be a new molecule or a new mixture of existing materials.

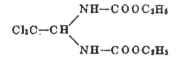

1. *Chloroaldiurethane* (see B. 42, 4067)

Preparation.—6 g. of urethane are dissolved in 5 g. of chloral and 3–4 drops of concentrated sulfuric acid are added thereto. The reaction mixture becomes warmed up and solidifies. It is added to water, then the small crystals separated out are sucked off and well washed out. Finally they are recrystallized from acetone. M. P. 172° C.

Formula.—

$$Cl_3C-CH \begin{cases} NH-COOC_2H_5 \\ NH-COOC_2H_5 \end{cases}$$

DDT Patent, U.S. Patent 2,329,074 (1943)

DDT, an insecticide, was claimed as a "Devitalizing Composition of Matter." DDT reduced malaria and typhus in some regions and won the Nobel Prize in Physiology or Medicine for Paul Muller, but it also raised environmental concerns.

Doogie Mouse, Princeton

New combinations of genes may be patented as a composition of matter. Neurobiologist Joe Z. Tsien patented the NR2B gene, which improved the memory of mice, known as the Doogie Mouse.

Two other notable compositions of matter:

- Tetrafluoroethylene Polymers, better known as Teflon, the nonstick coating for pans. As they say, "if nothing sticks to Teflon, how do they get it to stick to the pan?"[7]
- Nylon (U.S. Patent 2,071,250, 1937, styled as Linear Condensation Polymers), discovered by accident at DuPont.

Questions

1. Which composition of matter improved warfare and still funds a Peace Prize?

United States Patent Office.

ALFRED NOBEL, OF HAMBURG, GERMANY, ASSIGNOR TO JULIUS BAND-
MANN, OF SAN FRANCISCO, CALIFORNIA.

Letters Patent No. 78,317, dated May 26, 1868.

IMPROVED EXPLOSIVE COMPOUND.

The Schedule referred to in these Letters Patent and making part of the same.

TO ALL WHOM IT MAY CONCERN:

Be it known that I, ALFRED NOBEL, of the city of Hamburg, Germany, have invented a new and useful Composition of Matter, to wit, an Explosive Powder;

The nature of the invention consists in forming out of two ingredients long known, viz, the explosive substance nitro-glycerine, and an inexplosive porous substance, hereafter specified, a composition which, without losing the great explosive power of nitro-glycerine, is very much altered as to its explosive and other properties, being far more safe and convenient for transportation, storage, and use, than nitro-glycerine.

2. What "supermatter" is this?

Oct. 23, 1956   H. W. COOVER, JR   2,768,109

ALCOHOL–CATALYZED α–CYANOACRYLATE ADHESIVE COMPOSITIONS

Filed June 2, 1954

Fig. 1

UNSEASONED WOOD

α – CYANOACRYLATE ADHESIVE CONTAINING ALCOHOL

Answers

1. Dynamite.

Dynamite Demonstration

2. Super Glue.

Processes

Patented processes can be inventions or improvements on methods of (1) doing something, (2) using something, or (3) making something.[8]

Processes as inventions

The word "invention" evokes images of such things as the gramophone, telephone, or xylophone. But processes may also be patented. The first U.S. patent was for a "Method of making potash."[9] Pixar Animation Studios has a number of patents on technological methods for computer animation. Mark Pringle patented a method of making uniform potato chips (a.k.a. "crisps").[10]

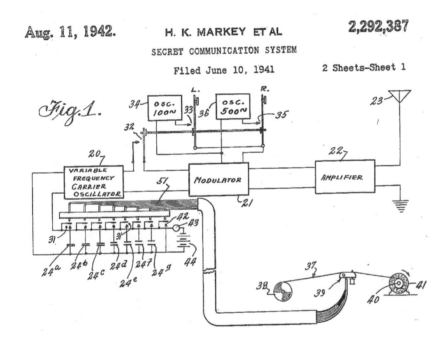

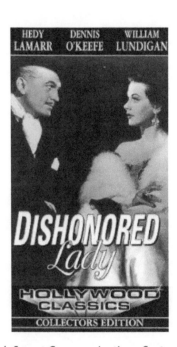

Hedy Lamarr's Secret Communications System

Movie star Hedy Lamarr rebutted any number of stereotypes during World War II by inventing a device to guide torpedoes by radio and hop frequencies to avoid jamming by enemy ships.[11] Her co-inventor, George Antheil, was a modernist composer, à la Stravinsky, whose works likewise feature frequency hopping. Meanwhile, frequency hopping has become a staple of cell phone technology, allowing many users on limited bandwidth. Note again how much shorter patent term is than copyright. If patent had a term of 95 years, cell phone bills would have a line item for a Hedy Lamarr license fee.

Kary Banks Mullis won the Nobel Prize for the polymerase chain reaction, which quickly creates millions of copies of a DNA molecule, permitting everything from DNA testing to the sequencing of the human genome (or other organisms).[12]

UNITED STATES PATENT OFFICE.

CHARLES GOODYEAR, OF NEW YORK, N.Y.

IMPROVEMENT IN INDIA-RUBBER FABRICS.

Specification forming part of Letters Patent No. 3,633, dated June 15, 1844.

To all whom it may concern:

Be it known that I, CHARLES GOODYEAR, of the city of New York, in the State of New York, have invented certain new and useful Improvements in the Manner of Preparing Fabrics of Caoutchouc or India-Rubber; and I do hereby declare that the following is a full and exact description thereof.

My principal improvement consists in the combining of sulphur and white lead with the india-rubber, and in the submitting of the compound thus formed to the action of heat at a regulated temperature, by which combination and exposure to heat it will be so far altered in its qualities as not to become softened by the action of the solar ray or of artificial heat calender-rollers, by which it may be brought into sheets of any required thickness; or it may be applied so as to adhere to the surface of cloth or of leather of various kinds. This mode of producing and of applying the sheet caoutchouc by means of rollers is well known to manufacturers. To destroy the odor of the sulphur in fabrics thus prepared, I wash the surface with a solution of potash, or with vinegar, or with a small portion of essential oil or other solvent of sulphur.

When the india-rubber is spread upon the firmer kinds of cloth or of leather it is subject to peel therefrom by a moderate degree of force, the gum letting go the fiber by which the two are held together. I have therefore devised

Patent on Vulcanizing Rubber

Goodyear's process for vulcanizing rubber made possible the era of the automobile and more—even though Goodyear never got much out of it. Process inventions loom large in every area, from software to business methods to manufacturing and scientific techniques.

Questions

Recognize these processes?

1.

L. PASTEUR.

Brewing Beer and Ale.

No. 135,245.   Patented Jan. 28, 1873.

2.

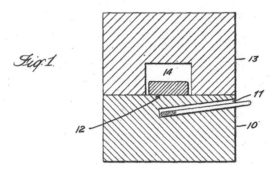

Answers

1. Pasteurization: Louis Pasteur's process of heating edibles to kill pathogens in food and drink.
2. Nuclear fission process.

Product and process together

An inventor can patent both process and product. For example:

Jan. 28, 1947.   R. T. JAMES   2,415,012

TOY AND PROCESS OF USE

Filed Aug. 21, 1946   3 Sheets—Sheet 2

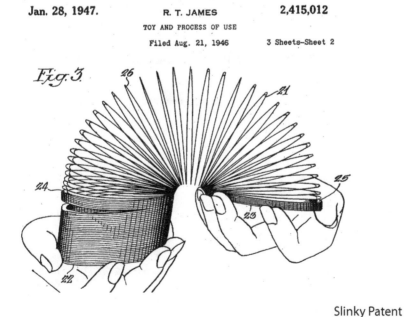

Slinky Patent

By patenting both a process and a product, the patent protects more broadly.

Processes as improvements

Not only pioneer inventions may be patented. Improvements, large or small, may be patented. Raymond McJohn, the great uncle of one of the authors of this book, patented an improved ironing board. (Not *the* ironing board—the Sears Tower, not the McJohn Tower, looms over Chicago, and it is not built in the shape of an ironing board.) Rather, he patented an ironing board with a safety feature.

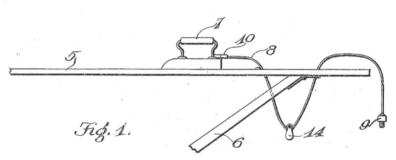

An Improvement Patent (Ironing Board Designed to Keep the Electrical Cord Safely Out of the Way, U.S. Patent No. 1693615, Granted to Raymond McJohn, 1928).

Long after the Wright Brothers, Paul MacCready patented an aircraft so light it could be powered by pedaling, the Gossamer Condor.

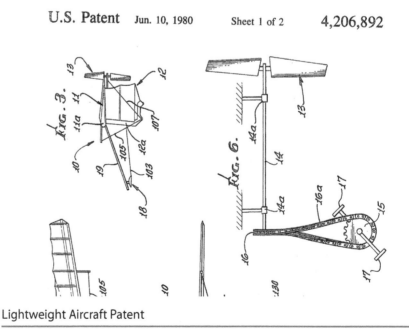

Lightweight Aircraft Patent

By adding multiple stages to rockets, Robert Goddard opened the path that later put things into space.

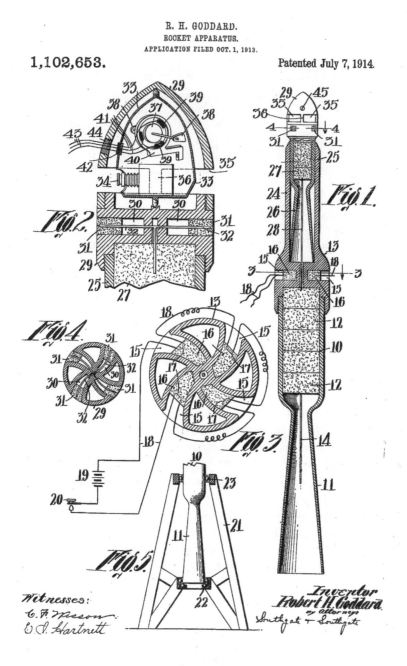

R. H. GODDARD.
ROCKET APPARATUS.
APPLICATION FILED OCT. 1, 1913.

1,102,653.

Patented July 7, 1914.

Witnesses:
C. F. Nesson
E. L. Hartnett

Inventor
Robert H. Goddard
by Attorney
Southgate & Southgate

What I do claim is:—
1. In a rocket apparatus, in combination, a primary rocket, comprising a combustion chamber and a firing tube, a secondary rocket mounted in said firing tube, and means for firing said secondary rocket when the explosive in the primary rocket is substantially consumed.

Rocket Apparatus, U.S. Patent No. 1102653 (1914)

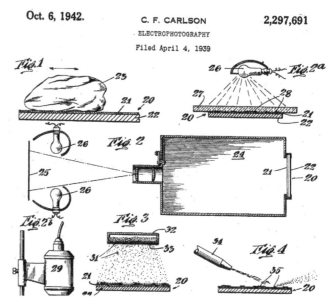

Oct. 6, 1942.　　C. F. CARLSON　　2,297,691
ELECTROPHOTOGRAPHY
Filed April 4, 1939

One of Many Xerox Patents

Chester Carlson invented "electrophotography." He and Xerox kept patenting improvements on the process, keeping control decades beyond the 17 years of the initial patent. Carlson had learned much working in the patent department at Bell Telephone, home for decades of the most potent patent portfolio ever.

Many notable inventions were explicitly styled as improvements. Anesthesia was "Improvement in Surgical Operations." Elisha Graves Otis's elevator was "Improvement in Hoisting Apparatus" (a hoist—that's why Harry Potter calls it a "lift"). Nicolaus August Otto's internal combustion engine was "Improvement in Gas-Motor Engines."[13]

Technological progress usually proceeds by improving or combining existing devices and processes, as opposed to making new ones. Most devices and processes bear little resemblance to the pioneer device. Compare today's car, computer, or phone to the early ones. Some devices from two centuries ago, however, have seen little improvement since, such as the safety pin, the paper clip, and the zipper.[14] Emerson and entrepreneurs notwithstanding, there has not been a better mousetrap since 1899.

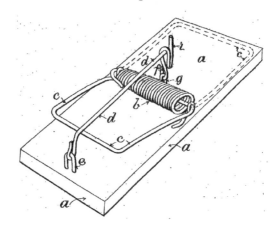

Great Britain Patent No. 13277 (1899)

What are these improvements each better known as? (The improvements themselves have since been improved upon.)

1.

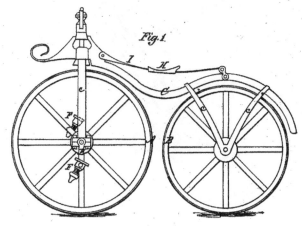

P. LALLEMENT.
VELOCIPEDE.

No. 59,915.   Patented Nov. 20, 1866.

Fig.1.

"Improvement in Velocipedes"

2.

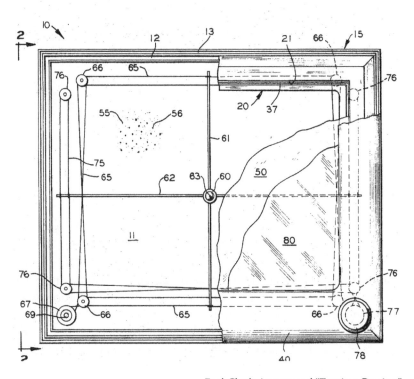

PATENTED SEP 25 1973   3,760,505

SHEET 1 OF 2

Earl Clark, Improved "Tracing Device"

3.

Ives W. Mc Gaffey,

Sweeping-Machine.

N° 91,145.   *Patented June 8, 1869.*

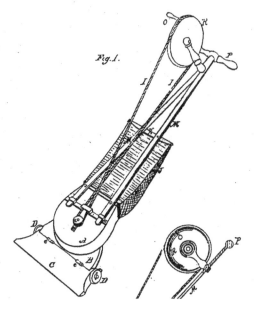

An Improved Sweeping Machine

Answers

1. Bicycle
2. Etch-a-Sketch
3. Vacuum cleaner

How long do patents last?

Engelbart invented and patented the computer mouse but never made much money from it. His patent was issued in 1970 and expired in 1987. By the time the computer mouse became a consumer product, Engelbart's patent had expired.

Utility patents (for invention) last about 17 years. Design patents last 14 years. That is a mere fraction of the copyright term (life plus 70 years for individuals, 95 years for works made for hire). If patent were like copyright, inventions from the 1920s would still be under patent protection. There is less cost to long copyright terms. Because of fair use and the nonprotection of ideas, others may still use the most important parts of copyrighted works. Picasso's and Dali's works may still be largely under copyright, but anyone can be a cubist or a surrealist. But technology builds more closely on previous work.

Few important copyrights have expired since the 1960s, because Congress has twice lengthened the term of copyright. By contrast, extremely valuable patents expire every year. Pharmaceuticals that sell for billions of dollars a year regularly reach the end of patent protection, at which point generic manufacturers sell their own version, usually for a fraction of the price.

Submarine patents

> When we grow stronger, then we'll make our claim:
> Till then, 'tis wisdom to conceal our meaning.

Shakespeare, Henry VI (III)

Patents last about 17 years. The law used to be that patents lasted 17 years. Therein lies a tale. The law until 1995 was that a patent lasted 17 years from when the claims were approved by the patent office, the necessary fees were paid, and the patent issued. Just *when* a patent issues is not controlled by the patent office. Rather, the inventor has a lot of control over how long it will take an application to become a patent.

The inventor can divide the application into separate applications, file continuing applications after claims are denied, and delay after claims are approved. Some inventors deliberately include at least one claim that will be denied. Such a zombie claim would be "I claim any machine or process that can solve the problem described in this application." That would prevent the patent from issuing until the person filing chose to abandon that claim.

Some inventors have been as clever in patenting as they were in inventing. An inventor can file an application on a device, bide time in the patent office while others in the industry develop and market devices in the field, then amend the claims to be sure they cover existing products and finally allow the patent to be issued—and then extract licensing fees from those selling the product. The best known example of this endeavor, known as "submarine patenting," involved a patent application filed in 1954 for an Automatic Measurement Apparatus that spawned divisional and continuation applications finally ending in the issuance in 1982 of a patent on a Scanning Apparatus and Method—whose claims then appeared to cover bar code technology. It would have been nice to have the patent on technology used in stores throughout the country.

The law was changed to address this issue. For applications filed after June 7, 1995, the term is 20 years from the date of filing, so delay simply reduces the patent term. An applicant may receive an extension by showing that the delay was due to the patent office or that there was delay by a governmental regulator.

Congress and the courts responded to submarine patents. Congress changed the patent term to 20 years from the date of application. So if the patent issued in 3 years, there would be a 17-year term. If it took longer, then the patent term would simply be shortened. That rule applied only to patents issued after the law changed. But as to existing patents, the courts applied a rule that an unexplained delay in receiving a patent could make the patent ineffective. That rule was applied in the bar code case, to the relief of many. Dealing with bureaucracies can be difficult, but an applicant that took 28 years to get through the patent office would have some explaining to do. His claim (really, his estate's claim, the patent struggle having long outlived its inventor) to have invented bar codes decades in the past would not be honored.

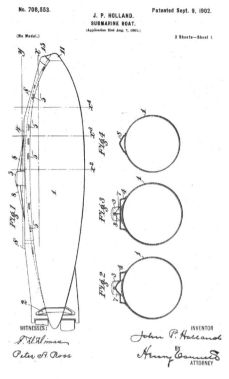

A Submarine Patent, Literally

Patents related to national security

There may be reasons why patents issue long after application. Inventions related to national security may be held under secrecy orders. An alert patent blogger noted that a patent on a method of jamming radar was filed in 1966, put under secrecy, and finally allowed to issue in 2011.[15]

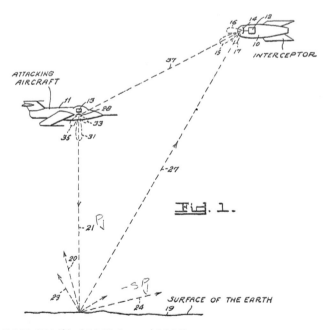

Patent No. 7,898,454 (filed 1966, issued 2011)

9

Nonpatentable: Humans, Nature, Ideas

Calvin: I'll probably go into genetic engineering and create new life forms.

Hobbes: You want to play God?

Calvin: Not exactly. God never patented his stuff.

Bill Watterson, "Calvin and Hobbes"

Patent law: Big tent

Patent law is not judgmental. Inventions may be patentable even if their use is questionable. The patented Juicy Whip device below, for example, serves only to deceive consumers. It dispenses beverages from the concealed container, not from the delicious bubbling display placed just above the dispenser. The device decreases the maintenance costs of cleaning the display bowl.

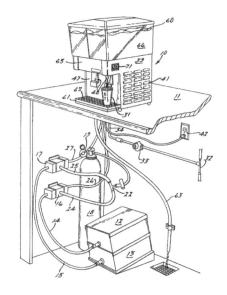

Patent on a Device to Deceive Buyers

Opinions differ about the social value of weapons such as the Gatling machine gun and gambling devices such as slot machines. But they may be patented.

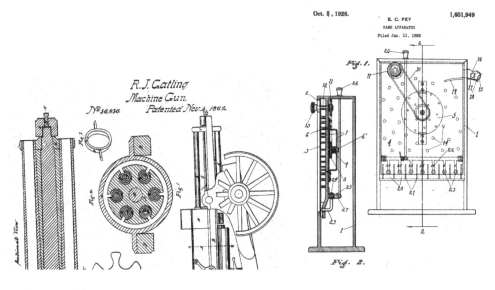

Gatling Gambling

A patent gives the inventor the right to exclude others from making, selling, and using the invention. It does not authorize the inventor to do those things. Rather, patent law leaves such issues as the legality of weapons, gambling devices, and consumer marketing devices to other areas of law (such as consumer protection law). Moreover, even if an inventor has a patent on a device, another patent may cover different aspects of the device, meaning the first inventor cannot make or sell the invention (unless the two come to an agreement). Likewise, patents may apply to devices used not for commercial purposes, but rather just to have fun.

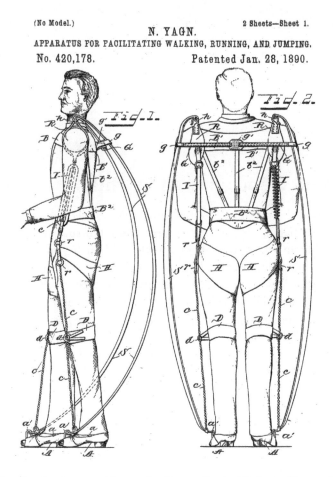

Gambolling Device Invented by, as the Patent Put It, "Nicholas Yagn, a Subject of the Emperor of Russia and Residing at St. Petersburg, Russia"

Patents are not limited to commercial inventions. Devices for playing, resting, wasting time, and pursuing dubious goals are patentable.

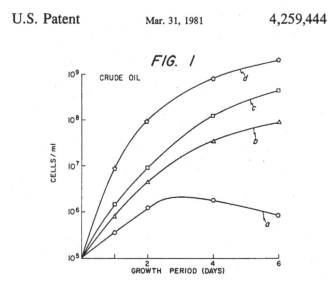

Crude Oil: An Acquired Taste

Living things are patentable. Chakrabarty genetically engineered a new bacterium that eagerly digested oil. Chakrabarty intended to sell the bacteria to countries in the Middle East, where it could transform their ample crude oil into protein that could be fed to livestock. The oil producers found more valuable markets for the oil. Chakrabarty then proposed using the bacterium to help clean up oil spills. The biotech industry has many valuable patents in the wake of Chakrabarty.[1]

No patents "encompassing a human"

One species of living things is not patentable: *Homo sapiens*. As of 2011, the patent law bars any patent "encompassing a human organism." The provision would bar patents on genetically altered humans and cloned humans, and likely chimeras combining humans and other species, not that any actual inventions were in sight as of 2011. One theoretically could read the law to bar patents on human stem technology or inventions affecting humans, such as medical treatments or nutritional programs. The vague phrase "encompassing a human organism" could encompass a hug. The U.S. Patent Office, however, soon stated its view that the exclusion is narrow and that it will continue to issue patents on drugs, treatments, and other human biotech inventions.

Questions

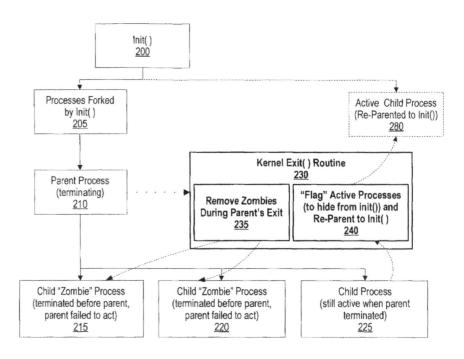

IBM's Patent Concerns Software Zombies

1. Are zombies (as in *Night of the Living Dead*) patentable?
2. Suppose someone managed to scrape together enough Neanderthal DNA to clone Fred Flintstone: patentable?

1. Are zombies human and so not patentable? Hmmm.

2. **U.S. Patent**   March 28, 1978   Sheet 3 of 3   4,081,183

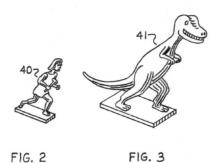

FIG. 2   FIG. 3

Drawing from Patent on Board Game Apparatus

A Neanderthal toy is patentable: Joseph Urban's game, described in the patent pictured above, "creates a novel pursuit situation in which one or more simulated Cavemen are pursued by a simulated Dinosaur along a path that contains specific hazards and shortcuts."

Is a Neanderthal a human? They were a different species than *Homo sapiens*, but our close genetic cousin. There is evidence of interbreeding between Neanderthals and modern humans.[2] Even if the bar against a patent "encompassing a human" does not apply, a Neanderthal is a product of nature (nature is discussed next) so would be unpatentable on that basis. A genetically altered Neanderthal is not a product of nature, but it could nevertheless be denied a patent as something "encompassing a human." All these situations are unlikely at present, but similar questions may actually arise as biotechnology develops.

Nature: Not patentable

Laws of nature are not patentable. The classic example: Einstein could not have patented his theories of relativity. They had little practical impact when he published them, by 1916. Not until decades later did practical applications exist—such as the atomic bomb in the 1940s and fine-tuning of GPS devices today. This example is instructive because it shows how broad patent protection for a law of nature could be.

The other classic example is Newton's law of gravity. It would be nice to have a patent on the law of gravity: everything from airlines to elevators would need a license to operate.

Even specific laws of nature are not patentable. The correlation between certain blood levels of a chemical and the likely effective dosage of the drug is a law of nature. A method for adjusting the drug dosage according to the correlation is not patentable.

Scientists seek to learn about laws of nature but also to develop practical applications. Many scientific achievements that won Nobel Prizes were also patented inventions, such as the transistor, the laser, MRI imaging, and the test for the HIV virus. All of those examples also involved battles over who should get scientific credit for the discovery and legal credit in the form of a patent. Some still disagree with the choices of the Nobel committee and the U.S. Patent and Trademark Office.

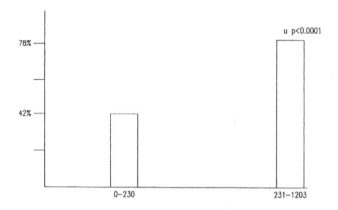

"Patent on Method of Treating IBD/Crohn's Disease and Related Conditions Wherein Drug Metabolite Levels in Host Blood Cells Determine Subsequent Dosage"

Question

Is the method of using ether as an anesthetic during surgery patentable?

Answer

The discovery that ether would effectively and safely render people unconscious for surgery was patented in broad terms in 1846. Today the claim would have to be more specific.

> It has never (to our knowledge) been known, until our discovery, that the inhalation of such vapors (particularly those of sulphuric ether) would produce insensibility to pain, or such a state of quiet of nervous action as to render a person or animal incapable to a great extent, if not entirely, of experiencing pain while under the action of the knife or other instrument of operation of a surgeon calculated to produce pain. This is our discovery, and the combining it with or applying it to any operation of surgery for the purpose of alleviating animal suffering, as well as of enabling a surgeon to conduct his operation with little or no struggling or muscular action of the patient, and with more certainty of success, constitutes our invention. The nervous quiet and

Jackson and Morton, Improvement in Surgical Operations, U.S. Patent 4,848

In 2012, the European Organization for Nuclear Research (CERN) discovered the Higgs boson, a long-sought subatomic particle. Without the Higgs boson, everything would zoom around at the speed of light, so it is quite useful. CERN may get Nobel Prizes or display their astrophysics skills on *Dancing with the Stars*, but it cannot patent the Higgs boson. Nor can Higgs. A product of nature is not patentable. The first person to strike gold or find a new species of flower cannot patent it. The discovery of the HIV virus was a milestone, but not patentable. However, a new product or process related to a natural phenomenon is patentable: a test for the HIV virus was patented jointly by French and American teams, after resolving a dispute about who discovered what.

Questions

1. Coolidge, a researcher at General Electric, improves Edison's light bulb. Rather than Edison's carbonized bamboo filament, Coolidge finds that tungsten heats more efficiently and lasts longer. Can GE patent a tungsten-filament light bulb, or would that simply be a patent on the superior qualities of tungsten?

2. *Elementary, my dear Watson*. Recall the Periodic Table of the Elements from chemistry class?

> Sir Humphry Davy
> Abominated gravy.
> He lived in the odium
> Of having discovered sodium.
> > Edmund Clerihew Bentley

Could someone patent an element?
Hint:

> There's antimony, arsenic, aluminum, selenium,
> And hydrogen and oxygen and nitrogen and rhenium,
> And nickel, neodymium, neptunium, germanium . . .
> And there may be many others, but they haven't been discovered.
> > Tom Lehrer, *The Elements*

(Note: AutoCorrect prefers "Nickelodeon®" to "nickel and neodymium.")

Answers

1.

UNITED STATES PATENT OFFICE.

WILLIAM D. COOLIDGE, OF SCHENECTADY, NEW YORK, ASSIGNOR TO GENERAL ELECTRIC COMPANY, A CORPORATION OF NEW YORK.

METAL FILAMENT.

| 1,026,383. | Specification of Letters Patent. | Patented May 14, 1912. |

No Drawing.   Application filed May 9, 1906.   Serial No. 316,006.

To all whom it may concern:
Be it known that I, WILLIAM DAVID COOL-IDGE, a citizen of the United States, residing at Schenectady, in the county of Sche-
terial to hold together the particles of finely divided tungsten. I consider it advisable 55 to add tungsten until the mixture contains about 55 per cent. by weight of tungsten.

Patent on Filaments of Tungsten and Other Metals

Coolidge cannot patent tungsten, and anyway, it was discovered long before Coolidge lived. But he can patent a useful device made of tungsten, leaving plenty of space for other inventors to work the tungsten space.

2.

United States Patent Office

3,156,523
Patented Nov. 10, 1964

1

3,156,523
ELEMENT 95 AND METHOD OF PRODUCING SAID ELEMENT
Glenn T. Seaborg, Chicago, Ill., assignor to the United States of America as represented by the United States Atomic Energy Commission
No Drawing.   Filed Aug. 23, 1946, Ser. No. 692,730
12 Claims. (Cl. 23—14.5)

The present invention relates to a new transuranic element. More particularly it is concerned with the transuranic element having atomic number 95, now known as americium having the symbol Am, isotopes thereof, compositions containing the same, and methods of producing and purifying said element and compositions thereof.

2

The fast neutrons are slowed down to approximately thermal energies by impacts with a moderator such as graphite or deuterium oxide, and the resulting slow neutrons (energies of 0–0.3 electron volt) are then absorbed by U^{235} to produce further fission and by U^{238} to produce U^{239} which decays through 93^{239} to 94^{239}. This self-sustaining chain reaction releases tremendous amounts of energy, primarily in the form of kinetic energy of the fission fragments. With such reactors the maximum reaction rate for steady state operation is determined by the maximum rate at which the heat of reaction can be removed. The rate of production of plutonium in such reactors may thus be equated, approximately, to the power output of the reactor, and amounts to about 0.9 gram of 94^{239} per megawatt day when operating with sufficient bombardment and aging times to permit total

A Patent Application on a New Element Would Not Be Granted Today

There was a patent on the first method to produce plutonium. But plutonium itself, it turns out, already existed in small quantities in nature. Patents were awarded to Glenn Seaborg on americium and curium in 1964, from a 1946 application. Perhaps during the Cold War the U.S. Patent and Trademark Office was amenable to patent applications from the U.S. Atomic Energy Commission. A new element that did not exist in nature was deemed patentable. Today, something as basic as an element would be deemed unpatentable as a product of nature.

UNITED STATES PATENT OFFICE.

JOKICHI TAKAMINE, OF NEW YORK, N. Y., ASSIGNOR TO PARKE, DAVIS & COMPANY, OF DETROIT, MICHIGAN, A CORPORATION OF MICHIGAN.

GLANDULAR COMPOUND AND PROCESS OF PRODUCING SAME.

945,638.   Specification of Letters Patent.   Patented Jan. 4, 1910.
No Drawing.   Application filed February 4, 1904. Serial No. 192,087. (Specimens.)

Adrenaline Patent

Stumble across a grizzly bear or go BASE jumping and your body will produce plenty of adrenaline. The form in which adrenaline appears in the body is a nonpatentable product of nature. An isolated and purified form of adrenaline, however, is considered something different: for "every practical purpose a new thing commercially and therapeutically."[3] So adrenaline was patented. Likewise, Merck Pharmaceutical patented vitamin B_{12}—in its isolated and purified state.

Genes?

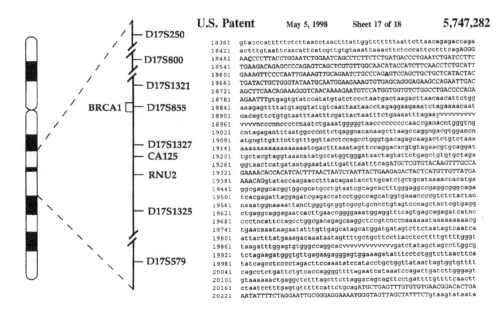

U.S. Patent 5,747,282 (17Q-Linked Breast and Ovarian Cancer Susceptibility Gene)

A question of law and philosophy: are human genes patentable? Not in the form in which they appear in the body, because that is a nonpatentable product of nature. Patents did issue (in similar fashion to the issue for adrenaline and vitamin B_{12}) for genes that have been identified and isolated. Myriad Genetics held patents on isolated *BRCA1* and *BRCA2* DNA, two mutations of human DNA linked to a higher probability of breast and ovarian cancer. The Supreme Court invalidated those patent claims, holding that isolated human DNA is not patentable. But the Court did hold that other related claims for cDNA, an artificial version that contains only the DNA sequences that actually code for genes, were patentable.[4] So genetic discoveries are subject to some patent protection. Do you think genes should be patentable?

Spurningar

1. Georges de Mestral looks through a microscope at the burrs that stick onto him when he is hiking. He fashions a material with similar little hooks—Velcro. Patentable or a natural phenomenon?

2. deCODE Genetics decoded the entire genomes of 1,795 Icelanders. deCODE found a genetic mutation in 1 percent of the population that provided protection against Alzheimer's disease by decreasing production of beta amyloid in the brain. The mutation is far rarer in other populations, indicating it arose in Iceland.[5] Patentable? Would a drug developed to mimic the effect of the mutation be patentable?

3. Which of the following Nobel Prize–winning advances are within patent subject matter?
   a. The universe is expanding at an accelerating rate (Perlmutter, Schmidt, and Riess, Physics, 2011)
   b. Buckypaper: one-atom thick layers of carbon (found using Sellotape®!) that have special properties (Smith, Geim, and Novoselov, Physics, 2010)
   c. Charge-coupled device (e.g., light sensor used in a digital camera instead of film) (Willard and Smith, Physics, 2009)
   d. The integrated circuit (Jack St. Clair Kilby, Physics, 2000)
   e. Discovery of the elements radium and polonium, and the isolation of radium (Marie Curie, Chemistry, 1911)

Svör

1. The fact that lots of tiny hooks will attach an item well is a nonpatentable natural phenomenon. But a human-made device to take advantage of the phenomenon is patentable.[6]
2. The genetic mutation as it appears in the body is an unpatentable natural phenomenon. The isolated gene with the mutation might be considered something different from what appears in nature, and so, patentable. A drug developed with the information would be patentable. A patent on the drug would not be a patent on the natural phenomenon (or the law of nature), because other drugs could be developed that would do similar things.
3. Here are the answers to the question about Nobel Prize–winning advances:
   a. This is a nonpatentable law of nature. What good would the patent be anyway? To sue someone living in an alternate universe for a court order to cease and desist expanding? The patent is a record of the achievement, but it pales next to the Nobel Prize.
   b. If that form also appears in nature, it is a nonpatentable natural phenomenon, even if no one knew about it until recently. But processes to isolate it or uses of it would be patentable.
   c.

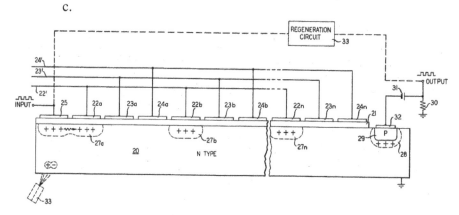

U.S. Patent 3858232 for Information Storage Devices

The CCD was developed in 1969 at Bell Labs, that font of both pure research and of the many patents in the fearsome Bell portfolio.

d. Here's the patent on the integrated circuit:

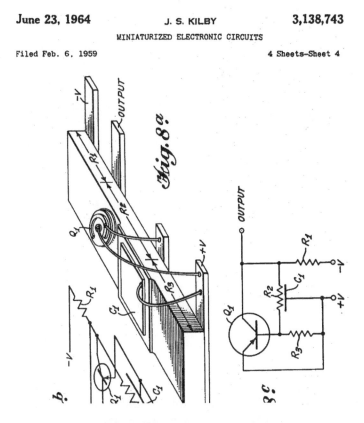

June 23, 1964   J. S. KILBY   3,138,743

MINIATURIZED ELECTRONIC CIRCUITS

Filed Feb. 6, 1959   4 Sheets—Sheet 4

Revolutionized Electronics (and Catches Nano Mice?)

e. Existing elements are nonpatentable phenomena of nature. Isolating an element in a form not normally found in nature *might* be considered patentable.

Abstract ideas not patentable

Patents are a trade-off. To encourage innovation, inventors get the right to prevent others from using or selling the invention without permission. If patents were too broad, the bargain would be one-sided. Imagine if in the early days of the commercial Internet, someone thought, "It would be nice to be able to search the whole Internet, rather than ping around to find things." If that general idea were patentable, then all search engines and other information locators would operate only with permission of that patent holder. Google has patents only on its concrete inventions, such as the Page Rank system, not the abstract ideas that drove its technical and business innovations. Claude Shannon's information theory revolutionized computing—but was not patentable.

The following table was the subject of a patent application.

Shown as the sum of powers of 2:

Decimal		2^3 (8)		2^2 (4)		2^1 (2)		2^0 (1)		Pure Binary
0	=	0	+	0	+	0	+	0	=	0000
1	=	0	+	0	+	0	+	2^0	=	0001
2	=	0	+	0	+	2^1	+	0	=	0010
3	=	0	+	0	+	2^1	+	2^0	=	0011
4	=	0	+	2^2	+	0	+	0	=	0100
5	=	0	+	2^2	+	0	+	2^0	=	0101
6	=	0	+	2^2	+	2^1	+	0	=	0110
7	=	0	+	2^2	+	2^1	+	2^0	=	0111
8	=	2^3	+	0	+	0	+	0	=	1000
9	=	2^3	+	0	+	0	+	2^0	=	1001
10	=	2^3	+	0	+	2^1	+	0	=	1010

A Nonpatentable Abstract Idea

A method of converting binary decimal numbers to pure binary numbers, using the table above, was held not patentable by the Supreme Court.[7] That's something computer programmers in many areas need to do. Patenting such a basic process would be like patenting the idea of a search engine.

The idea of hedging risk by selling derivatives, such as a farmer selling corn futures to guard against a drop in crop prices, likewise was too abstract to be patented.[8]

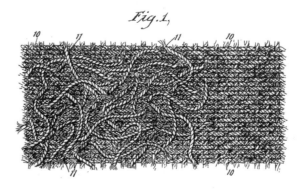

R. W. TULLY.
ARTIFICIAL HEDGE, LAWN, AND SIMILAR ARTICLE.
APPLICATION FILED OCT. 16, 1918.
1,386,450. Patented Aug. 2, 1921.

Fig. 1,

A Different Kind of Hedge

Although an abstract idea is not patentable, an *application* of that abstract idea is. For example, a patent was proper on a way of using an abstract method in a specific manner to manufacture rubber.[9]

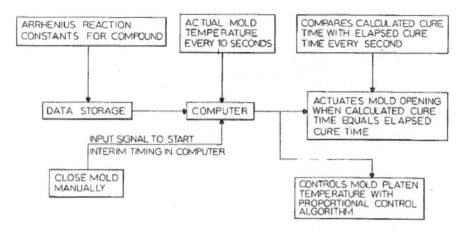

Patentable Application of Abstract Idea

The no-patents-on-abstract-ideas rule comes up most in software (such as the binary conversion method) and business ideas (such as using derivatives to hedge risk). There certainly are other areas in which people produce abstract ideas, such as mathematics, philosophy, economics, and literary theory. But no one tries to patent those until they have put them into some useful form, like software or a business method.

Question

Alvin Roth won the 2012 Nobel Prize in Economics for his work on matching theory.[10] Roth has applied his theory to designing "match day," which pairs medical students and residency programs, and to designing a program that assigns New York elementary school students to colleges based on their list of preferred schools.[11] He also designed a program to arrange kidney swaps. Say someone needing a kidney transplant has a relative willing to donate, but that person has the wrong blood type. By arranging things with other incompatible pairs, the recipients may be matched with compatible donors. The operations will be simultaneous because people can't sign legally binding contracts to donate kidneys (remember *The Merchant of Venice*). Could Roth have patented his theories? Could he have patented its applications?

Answer

Roth's theories are nonpatentable abstract ideas. Concrete applications may be patentable. Others have patented inventions applying Roth's work, as shown in the patent reproduced here.

Fig. 2
OBJECT STRUCTURE

Object Trader

- Personal_Characteristics

- Active Bid_Baskets: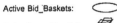

- Active Items_For_Sale:

- Transaction_History
- Active_Contractual_Agreements
- Reputation_and_Feedback

Object Item_For_Sale
- Trader Seller:

- Item_Characteristics
- Bidding_Information
- Bid_History

Budish System and Method for Conducting Electronic Commerce, U.S. Patent 7398229

New, Useful, and Nonobvious

Competing inventors: First to file
or make public

On February 14, 1876, Alexander Graham Bell and Elisha Gray each filed patent applications claiming the telephone. The patent office determined, some say influenced by payment from Bell's lawyers, that Bell was entitled to the patent. Bell became one of America's most famous inventors, and the Bell Telephone Company spawned the telecommunications industry and Bell Labs, responsible for the transistor, the discovery of cosmic microwave background radiation, information theory, and digital cameras, among other things. The Bell patent is likely the most valuable patent in history.[1]

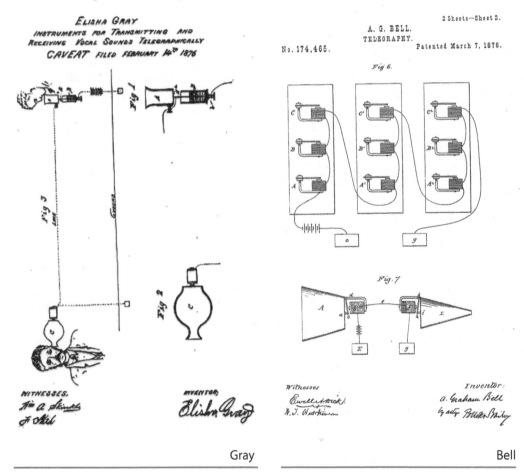

Gray   Bell

The old new

First to Claim: Oklahoma Land Rush 1889 (Sport of Would-Be Kings)

If two patent applications claim the same invention, the patent goes to the first inventor to file. One twist: if one inventor previously made the invention public within the last year, she takes the patent. The first-to-file rule itself is new. Until 2013 the United States used a first-to-invent system. The first person to invent something was entitled to a patent even if someone else filed an application first. If two applications covered the same invention, the patent office would hold an interference proceeding, not unlike a trial.

Most countries use a first-to-file system. To harmonize with the rest of the world (because an applicant often files for patents in many countries), the United States switched to a first-to-file system, effective for applications filed after March 16, 2013. First-to-file is simpler and more predictable. If two applications cover the same invention, no complex trial is necessary. The patent goes to the first application filed.

Some in the United States resisted switching to such a system. There was concern that first-to-file favored corporations over individual inventors. Indeed, with more sophistication about patent law and more resources, corporations are likely to win the race to the patent office. Others argued, though, that first-to-invent was no better for small inventors because corporations typically were more successful in interference proceedings for the same reasons—more experience with patent law and more resources. First-to-file is cheaper, so individual inventors need to invest less to protect their rights. Certainly under the old system there were many individual inventors whose rights were trampled by others who patented the invention: the vacuum cleaner was invented not by Hoover but by Murray Spangler; the sewing machine not by Singer but by Howe.[2] Which system do you think is more fair?

New

To be patentable, an invention must be new—or "novel," in patent jargon— at the time of the patent application. An applicant is not entitled to a patent if the invention is already in the public domain: patented, published, or in public use. If someone reinvents the wheel, he cannot patent it. The strong rights of a patent are to be given only to true innovators. The novelty requirement also encourages inventors to move quickly to seek a patent and so disclose their technology to the world.

The public disclosure could come from the inventor. Suppose a chemist publishes an article about a super-strong glue that will not stick to skin, but she does not file a patent application for a couple of years. Her invention would not be novel at the time of filing, and so it would not be patentable. The disclosure can come from someone else. Suppose an alchemist develops a process for turning lead into gold and quickly files his patent application, but someone else comes up with the same process and publishes it earlier.

The novelty requirement in the United States has a little give to it. If the inventor caused the invention to become public, there is a one-year grace period. The inventor may still receive a patent if she files within one year of her public disclosure. The grace period allows, for example, scientists to make discoveries public, so they need not remain mum until the patent application is filed. It also gives a little leeway to inventors less savvy about patent law—who could otherwise work hard on solving a problem, find a great invention, and then lose patent rights by yelling "Eureka!" through the streets, like Archimedes. The U.S. system effectively gives the patent to the first to file or disclose.

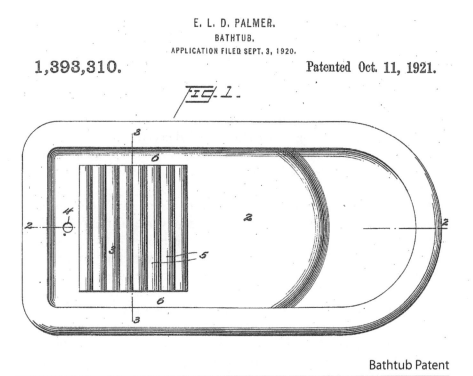

Bathtub Patent

Public use

Leader Technologies applied in 2003 for a patent, granted in 2006, on a software system that enabled users to collaborate and communicate through boards accessible through an Internet browser. Users could use the online boards to coordinate groups, share pictures and videos, express opinions, or chat with other users. Facebook, launched in 2004 as Thefacebook, provided a system that let users do similar things. Leader sued for Facebook for patent infringement.

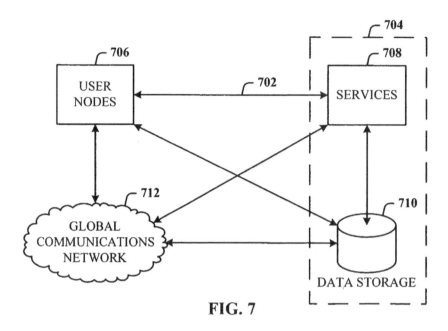

FIG. 7

Leader Technologies' U.S. Patent 7,139,761 ("Dynamic Association of Electronically Stored Information with Iterative Workflow Changes")

It would be nice to own the rights to the technology that Facebook depends on. But the courts held that Leader Technologies' patent was invalid. Leader's patented technology had been embodied in early versions of its software, and put into commercial use more than a year before Leader filed its patent application. Thus Leader's patent was invalid because the invention was not new.[3]

Low-tech inventions may also run afoul of the novelty requirement. A typist developed Liquid Paper, a little container of white enamel paint with a little paintbrush, used for easily correcting mistakes back in the age of type-writers. But little containers of enamel paint along with little paintbrushes had long been sold for other purposes, such as painting little figurines, so she could not have received a patent on it. However, she may well have been entitled to a process patent. A new use for an old product can be patented as a new process, and the process of using paint to correct mistakes in typing may have been new. But such a patent would be difficult to enforce—getting lots of tiny judgments against typists as opposed to being able to sue manufacturers of a product. She did not seek a patent but nevertheless marketed the product with great success, reminding us that innumerable successful products were never patented.

An inventor may also be barred by the work of others. Eolas held a patent that covered basic Internet technology, the browser. A number of companies bought licenses from Eolas or settled litigation (most notably, Microsoft). In 2012, however, a jury found the patent invalid. Experts, including Tim Berners Lee, the inventor of the World Wide Web, had testified at trial that similar technology, such as the Viola browser, had been in public use before Eolas developed its browser.

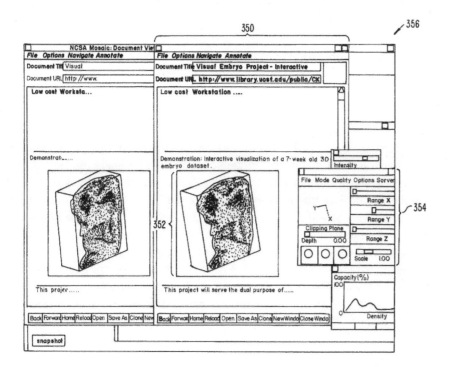

U.S. Patent 5,838,906 (Eolas Browser Patent)

Question

Only new things can be patented. Tricycles and lawnmowers were not new in the 1980s. So how could they have been patented?

U.S. Patent   Jun. 26, 1984   **4,455,816**

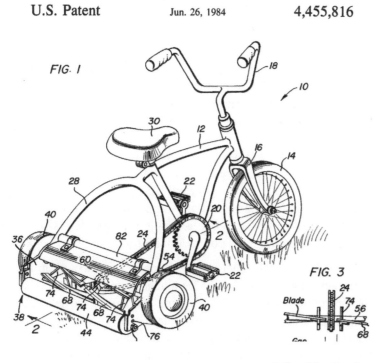

A Combination Invention

A new combination of existing elements is new. All patented inventions involve combining known elements.

Published

Hall, a chemist, developed a handy enzyme and applied for a U.S. patent. Unbeknownst to Hall, a German graduate student had already developed the same enzyme and described it in his doctoral thesis, with the catchy title "1,4-a-Glucanglukohydrolase ein amylotylisches Enzym." The thesis had been filed and cataloged in the Freiburg University library. Because the enzyme had been so published, Hall was not entitled to a patent. It did not matter that Hall had been unaware of the doctoral thesis or that the thesis was on another continent in another language. Nor did it matter if no one but the patent examiner had ever read the thesis. That which has been published is not patentable. Patents have been barred by such publications as poster boards at scientific conferences, any manner of online postings, and handouts at public meetings. A Danish inventor sought a patent on raising sunken ships by inflating devices inside. But the application was itself sunk, reportedly, by a cartoon in which Donald Duck filled a sunken boat with ping-pong balls to raise it.[4]

Question

Inventors filed a clinical trial plan with the Food and Drug Administration to test a new drug to treat strokes. The plan was published as a result of a Freedom of Information Act (FOIA) request. A couple of years later, the trial showed the drug to be effective. The inventors filed their patent application. Problem?

Answer

Too late to file. The inventor loses the right to a patent if the application is not filed before someone else publishes or less than a year after the inventor herself publishes.[5]

Why didn't the inventor file the patent application earlier? Before the trial, she may not have had data to show the invention was effective. So an inventor may be stuck between a rock (must show usefulness, discussed shortly, at filing) and a hard place (need to file no later than a year after publishing).

Inventors may be able to prevent publication. They may keep their scientists from publishing journal articles. Also, FOIA requests for confidential commercial information may be denied.

Igor Sikorsky flipped the switch and patented the helicopter in the 1930s. Leonardo da Vinci had drawn up plans for a helicopter almost 500 years earlier. The publication of da Vinci's sketch, however, did not make the helicopter unpatentable. Da Vinci described an idea to make air flow down, making the device go up, but his sketch lacked any number of technical details to make a functioning helicopter.[6] Sikorsky, among many other mechanical solutions, used propellers on top and behind. With just a propeller on top, the helicopter itself would spin around, as we know from Newton's Third Law of Motion from 1687.

Da Vinci's Arial-Screw (1480)

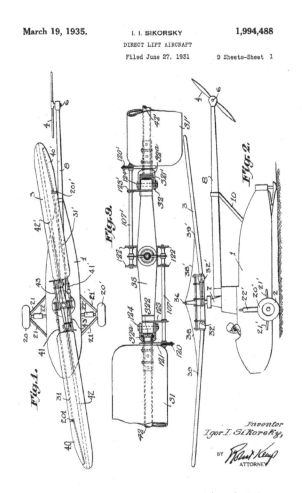

March 19, 1935.   I. I. SIKORSKY   1,994,488

DIRECT LIFT AIRCRAFT

Filed June 27, 1931   9 Sheets-Sheet 1

Inventor
Igor I. Sikorsky,

BY
ATTORNEY

Sikorsky's Direct Lift Aircraft (1931)

Only the inventor is entitled to apply for a patent. Sperry Rand held a patent on ENIAC, the first digital computer. But when Sperry Rand sued Honeywell for infringement, the court ruled that ENIAC was derived from the true inventor, Atanasoff, and from prior publication by von Neumann. The patent was invalid and could not be enforced, meaning that Sperry Rand could not control the first generation of commercial computers.[7] Likewise, an anthropologist that learns of a traditional medicine from an indigenous group is not entitled to a patent.

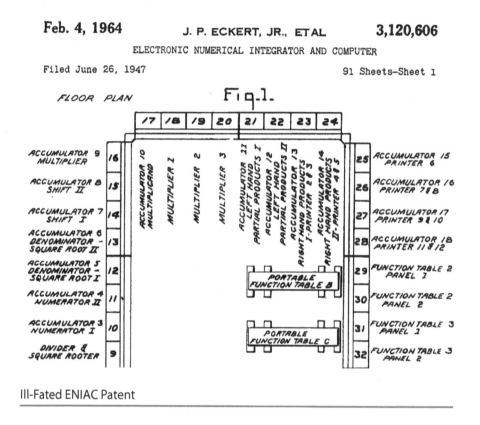

Ill-Fated ENIAC Patent

Question

Some Californians read in a newspaper that Australian kids twirl bamboo hoops around their waists for fun. The Californians develop the Hula Hoop. Patentable?

Answer

The Californians did not invent the device; rather, they learned of it from another, so they are not entitled to a patent. To double down, the invention was not new; it had been published. The unpatented Hula Hoop nevertheless made millions for the Californians. Others could copy their device but not use their trademark, Hula Hoop®.

Useful

To be patentable, an invention must be useful.[8] An application for a patent on a new walking stick need not show that it is better than existing walking sticks in any way but simply that it works. An invention may be patentable even if it is unsafe, inefficient, overly expensive, or works only inconsistently. An inventor may seek a patent long before the invention is ready for commercial release. A patent was issued for a transistor that worked to amplify current but was commercially impractical. It was made from germanium, a rare, costly element that malfunctioned at low temperatures. But the patent was valid, although there is now no Germanium Valley in California. For most areas of technology, the requirement of usefulness is easily met. A new machine, circuit, or piece of software will likely function, whether or not it is better than existing technology.

The invention need only be somehow useful. The inventor need not identify all its possible uses. Inventors notoriously fail to envision the future. Bell thought the telephone would be primarily useful for listening to symphonies from other cities. He did not foresee its primary use of social networking (history repeated itself with the Internet). Edison did not see a use for the movie camera beyond recording family events. His one film was a boxing match between cats. Although Edison did not anticipate movies, he did prefigure a principal use of the Internet: cute cat videos.

In some areas of technology, the requirement of present usefulness has bite. In chemistry and biotechnology, researchers often have a new molecule or other innovation with potential but no proven practical application. A new steroid molecule was thought to have potential to treat inflammation, but without further testing, it was not shown to be useful. A patent "is not a hunting license."[9] If the inventor files too soon, the requirement of usefulness will not be met. But if the inventor waits to have proof of usefulness, someone else may publish or file first. Similarly, if the inventor has disclosed the invention (such as by publishing a paper in a scientific journal, or announcing it on a poster board at a scientific conference), then he will lose the right to a patent. The usefulness requirement and the novelty requirement combined may whipsaw an inventor.

Science fiction features all kinds of fantastic devices. *The Hitchhiker's Guide to the Galaxy* described a computer, Deep Thought, which computed the answer to the Ultimate Question of Life, the Universe, and Everything (the answer: 42). The book also described an "improbability drive" that jumped ships through space, and babel fish, which sit in your ear and translate languages. None of these are patentable, for lack of present usefulness—they do not work outside the imagination of the writer, yet.

Many a patent application for a time machine, perpetual motion device, or cold fusion power generator has foundered on the usefulness requirement. Unless the inventor can describe how to make and use the device, usefulness is not shown. Whether such impossible devices are patentable or not is not too important. A patent on a nonworking time machine would not make any difference. Nobody could infringe the patent. But denying such patents protects the credibility of the patent office.

1. An invention has to work to meet the usefulness requirement. So how did someone patent a Ouija board, the device that spirits use to spell out messages during a séance?

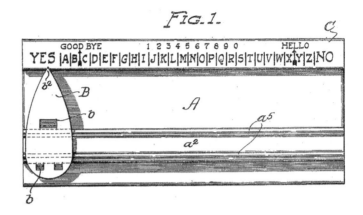

Ouija Board Patent

2. A device to allow one to watch television while driving a car: useful enough to be patented?
3. Would an exploding pen be useful?
4. Scientists learn how to make light bend at the nano level, managing to hide a few molecules from view. In theory, this could lead to an invisibility cloak like Harry Potter's. Patentable?
5. Scientists create a new element, albeit one that can last only a small fraction of a second. Patentable?
6. A patent application claims a spacecraft powered by antigravity, creating a space-time curvature anomaly that propels the vehicle at the speed of light. Said vehicle has not been built. Leading scientists find the description incredible (not in a good way). Can it be patented?
7. Can a drug be patented that has been shown to be effective in animals but not yet in humans?

Answers

1. The inventor's description of the Ouija board shows its usefulness.

> My invention relates to an improved structure of ouija boards.
> The ouija board, as is well known, is a device designed to permit human beings to give expression to subconscious thoughts induced by complete muscular and mental relaxation. Its purposes and principles of

2. This inventor also explained the usefulness of his device, though not successfully—patent denied.

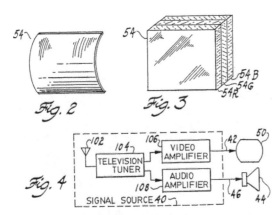

bile. This does not interfere with the visual task of operating the automobile. On the contrary, it has been found that in some occasions this display actually enhances the attention of the driver to the visual task of operating the automobile.

Autovision won a 1993 IgNobel Prize in Visionary Technology.[10] The IgNobels are awarded by the magazine *Annals of Improbable Research* for "achievements that cannot or should not be reproduced."

3. Yes, the exploding pen was deemed useful.

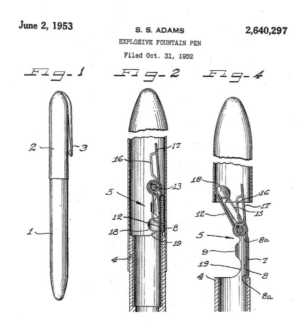

June 2, 1953   S. S. ADAMS   2,640,297

EXPLOSIVE FOUNTAIN PEN

Filed Oct. 31, 1952

Explosive Fountain Pen Patent

So silly inventions can be useful. The same inventor patented the trick pencil and the joy buzzer.

4. A promising scientific discovery is not patentable if it is not yet useful. When they build an invisibility cloak, that would be patentable.

5. As discussed in the last chapter, a new element might be patentable. But if it lasted just a moment, there may be no practical usefulness. On the other hand, if a new element served as a super power supply (think of palladium, the new element created by Tony Stark in *Iron Man*), then it would have usefulness.
6. It should not have been patented—but it was.[11]

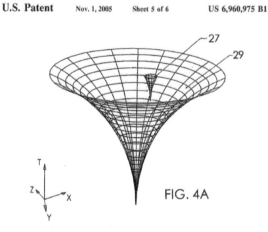

Space Vehicle Propelled by Pressure of Inflationary Vacuum State

The examiners at the U.S. Patent and Trademark Office are only human; they sometimes take the path of least resistance when faced with a complex claimed invention.

7. An invention must be useful—but need not be fully ready for the market. If a drug shows some effectiveness in tests, whether in animals, test tubes, or people, it has enough usefulness to be patented. Whether it can be marketed will depend on another government agency, the Food and Drug Administration, whose standards are higher.

Nonobvious

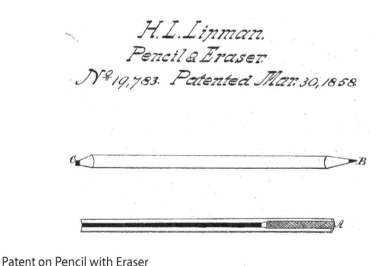

Patent on Pencil with Eraser

Years after the first pencil with an eraser attached was sold, Linman procured a patent on a pencil with an eraser attached, but inside the pencil. But the patent was invalid, as a trivial difference from existing pencil technology.[12] Even if an invention is new and useful, the inventor is not entitled to a patent if the invention would have been obvious to a typical worker in that area of technology. The bar to obvious inventions prevents patents on merely incremental inventions that would have been made anyway.

What does it mean for something to be obvious? We do not judge it from the actual process that the inventor followed. Many inventions were lucky. Standing by a magnetotron, Percy L. Spencer noticed the bar of chocolate in his pocket melted. Soon he patented the microwave oven.[13] A chemist noticed a sweet taste on rolls he had touched, which led to blockbuster artificial sugar saccharine. Charles Goodyear spilled rubber and some chemicals on a stove and thereby discovered the process dubbed vulcanization, which made rubber into the sturdy standby it is now. Nylon and Super Glue were also discovered when curious chemists investigated accidental concoctions. Gorilla Glass resulted from a Corning Glass furnace malfunctioning and overheating the material. Airy Ivory soap was supposedly first made by accident, when a factory worker neglected his job and the soap mixture was aerated.[14]

Many inventors were actually trying to make something else. British engineers seeking to develop a radiation weapon found that their electromagnetic waves did not knock planes down but could help detect them: radar. Chemists seeking a substitute for rubber during World War II instead invented Silly Putty. Leland Stanford, a lover of horse racing, the so-called sport of kings, hired photographer Eadweard Muybridge to settle a bet on whether a galloping horse's feet ever all left the ground at once. Muybridge's rapid series of photos became the first motion picture. There is no record on whether Stanford bet that movies would catch on.

Muybridge, *Galloping Horse* (1878)

Of course, many inventions are the result of research programs. Edison's crew tested thousands of materials before finding that carbonized bamboo worked as a long-lasting filament for incandescent light bulbs (nowadays tungsten is used). The transistor was the culmination of years of research at Bell Labs seeking something better than vacuum tubes. But obviousness is not judged by how the inventor came up with the invention.

Rather, the question is how different the invention is from existing technology. This approach helps avoid hindsight. Would the invention be obvious to someone working in that area of technology if that person knew of the relevant technology disclosed by publications, patents, and public uses? One also can look for indirect evidence of nonobviousness, such as commercial success (licensing or adoption as industry standard), solving a long-standing problem, unexpected results, or copying by others.

Questions

1. Crocs patented foam sandals with foam straps, which sold like wildfire, leading others to make knockoffs. When Crocs sued, alleged infringers argued that the invention was unpatentably obvious: straps had been used on shoes, and foam had been used in shoes. But sandal makers had not used foam straps, because foam is not flexible. Would foam straps have been obvious?

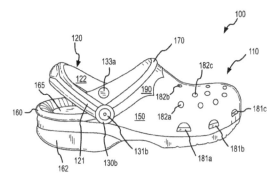

2. Western Union used telephones and faxes to wire money around the world. It was the first in its industry to move to online fund transfers, although other industries were online. Can Western Union patent the method and maroon Moneygram offline?

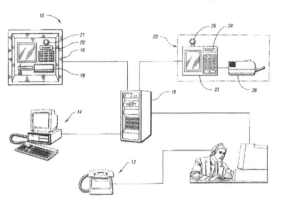

Online Wire Transfer Patent

3. Although da Vinci drew helicopters, Sikorsky's helicopter was nevertheless new because da Vinci did not show how to make one. But did da Vinci make the helicopter obvious and so render Sikorsky's helicopter unpatentable?

4. In 1936, the Sun Valley ski resort sought a better way to get skiers to the top of the mountain than towing them to the hilltop. James Curran had spent time in warmer climes and seen conveyor belts with hooks to carry bunches of bananas onto ships. He adapted the device to transport bundled-up humans up hills—and soon patented the Aerial Ski Tramway. Obvious?

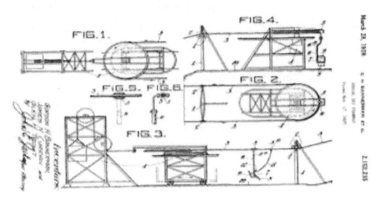

Ski Lift Patent

5. Around 1900, wristwatches were exclusively designed for women. Cartier designed a wristwatch for men to help a balloonist friend who had trouble piloting the craft while reaching for his pocket watch. Patentable?

See the invention? It's on the wrist of the zeppelin pilot.

Answers

1. Previous work in the field had discouraged the use of foam straps, meaning this invention was apparently not obvious. Once sold, the

sandals with foam straps were hugely popular, likewise suggesting that the product was not obvious.[15]

2. Online fund transfers were new—but obvious. Once the Internet was commercialized, it was obvious to simply adapt the fund transfer system using telephones and faxes to an Internet implementation.[16]

3. The differences between Sikorsky's invention and the prior art (da Vinci's helicopter sketches, along with any other relevant technology) were considerable. Sikorsky used a combination of existing elements—but in a complex, unprecedented combination. Moreover, had the helicopter been obvious, surely someone would have made the device in the intervening 500 years. A general idea does not make a device obvious, especially where the devil is in the details.

4. The technology was adapted and transformed to another area. Converting tropical banana conveyors would not be obvious to cold-climate ski-hill developers.

5. A wristwatch for a man was new but obvious, given the existence of wristwatches for women. But Cartier became the premier brand, *sans* patent.

Getting a Patent

Patent prosecution

Under the first patent law of 1790, patent applications were considered by a committee composed of the secretary of state (Thomas Jefferson), the secretary of war, and the attorney general. Jefferson quickly found that even with few patent applications, the task took up time better used in diplomacy with other countries. Congress created the U.S. Patent and Trademark Office (USPTO). Later cabinet members such as Henry Kissinger and Robert Kennedy were spared the task of considering patent applications. With over 8 million U.S. patents issued (and some 2 million in force now), that would have been quite a job.

Albert Einstein, Patent Examiner

Hong Kong Science Museum, http://hk.science.museum

The patent office directs each patent application to an examiner in the appropriate "art unit."[1] The examiner specializes in an area of technology. Some units are quite specialized. Chemistry 530.354 handles gelatin chemicals, which could include Jell-O and gelignite. Others are a little broader: Unit 2914: Arms, Pyrotechnics, Hunting and Fishing Equipment. The examiner spends a few dozen hours over the next few years deciding if the application merits a patent. First she does a patent search to help determine whether the invention is new and nonobvious. In theory, she would search every relevant publication, patent, or public use in the world up to the time of the patent application. In practice, of course, that is not possible, and although examiners have access to technical libraries, databases, and the Internet, they tend to focus on previous patents in related technology.

Applicants are not required to search for prior art before submitting an application, but must disclose in the application relevant prior art that they know of. The examiner examines the application in light of the prior art and issues a preliminary decision to approve or reject the "claims," the part of the application that defines the invention. The examiner may reject claims on the grounds that the claimed invention is not new, useful, or nonobvious, or because the applicant has not disclosed enough information, or for other reasons.

If examiners reject a claimed invention as unpatentable, which they often do on first pass, applicants can choose to amend the claims or convince examiners to change the decision. The applicant may request a face-to-face meeting, or appeal the ruling within the USPTO and then appeal to the courts. The applicant may also respond to rejection by filing a request to continue the application. In fact, there is never an absolutely final rejection because the applicant may always file some kind of continuation or divisional application or appeal the rejection.

Applicants often amend rejected claims to meet the objections of the examiner. In the majority of cases, after some to and fro, the examiner will approve the application and a patent will issue. This may take several years to occur. The USPTO is busy.

There are systemic problems in the U.S. patent system. The USPTO receives around half a million applications a year and now is so backed up that it typically takes three to five years to decide whether to issue a patent. Unlike many countries, there is ordinarily no opposition process in the United States while the application is pending, so the decision depends on input only from the applicant and the overburdened examiner. The easiest path may be for the examiner to grant the application, avoiding disputes with the applicant and clearing the caseload a little.[2]

U.S. Patent and Trademark Office

http://siarchives.si.edu/oldsite/history/exhibits/baird/patoff1.jpg

"The name of the game is the claim"[3]

Key to the patent application and to the patent, if it issues, are the claims. The inventor must *claim* her invention by describing it in words as a product or process and by listing its elements. Examiners look to the claims to see if the inventor is entitled to a patent by claiming a new, useful, and nonobvious invention. They look to the claims to see what the patent covers: what products or processes would infringe the patent.

The classic example of a claim: Imagine the first person who invented not the pencil nor the eraser, but the combination of a pencil with an eraser attached. That would have been a patentable invention. It was new (if that combination had never been publicly done before), useful (saved the pencil user the trouble of locating an eraser every time a mistake must be fixed), and nonobvious (if no one had done such a useful thing before, it must not have been obvious). The claim could have read:

> I claim: A pencil having an eraser fastened to one end.[4]

A patent claim may be broad or narrow—and that affects (1) whether the claim is likely to be a patentable and (2) if granted, how strong the patent will be to enforce. A narrow claim is easier to patent but gives less power to enforce. A broad claim is harder to patent but can be more strongly enforced. To make this idea concrete, imagine our inventor is not sure if pencils with erasers attached are new. Her claim will be invalid if a German grad student previously published a thesis on "ein Bleistift mit einem Radiergummi angebracht" ("a pencil with an eraser attached," according to Google Translate). Our inventor could decide to claim something definitely new by adding two other useful features that she came up with:

> I claim: A pencil having a *square* eraser fastened to one end *and rubber grips around the pencil.*

It is extra useful for the eraser to be square because it will not roll off the table as easily, and the grips help with the fine points of drawing.

Our inventor would be correct in thinking that her narrower claim would be easier to patent. The more elements we have, making the claimed invention more specific and so making the claim narrower, the less likely it is that something like that would be already publicly disclosed—or that it would have been obvious to make the invention.

But the trade-off would be that the patent is not as strong. It would not be infringed by a mere pencil with an eraser attached but only by a pencil with a square eraser and grips. Our inventor would have less market power as a result and would have to stand by helplessly while others sold her basic invention.

Patent law to the rescue: our inventor need not choose between broad and narrow claims. Rather, applicants may, and most do, submit multiple claims. An independent claim may be followed by dependent claims. For example:

> Claim 1
>
> I claim: A writing implement, comprising a pencil having an eraser fastened to one end.

Claim 2

The pencil of Claim 1, wherein the eraser is square.

Claim 3

The pencil of Claim 2, wherein the pencil has rubber grips around it.

This recursive structure, with each claim defined by the preceding claims, allows the inventor to start with a broad claim and add increasingly narrow claims.

Claim drafting is a skilled art, so most inventors hire a lawyer or patent agent to represent them. An inventor can draft and file the application, but there are many traps for the unwary. A classic DIY error is to draft a claim that lovingly describes the inventor's cherished invention in great detail. A patent may well issue, but no one would be guilty of infringing the patent unless his product or process also had every single one of those details.

In days of yore, patent claims were easier to read. An inventor could describe the invention in the application (including drawings), and submit a short claim essentially saying "I claim that." Bell's telephone patent simply claims "A System of telegraphy in which the receiver is set in vibration by the employment of undulatory currents of electricity, substantially as set forth."

The law now requires the inventor to distinctly claim the invention, element by element. Engelbart's patent on the computer mouse has a claim of typical length and impenetrable technical terms:

I claim:

1. In a display system controlled by a computer whereby the display is alterable in accordance with signals delivered to said computer which indicate positions on said display and changes desired to be made therein, the improvement in a position indicating control apparatus which is movable over a surface to provide position indications corresponding to positions on said display comprising:
   a. Housing;
   b. First position wheel rotatably mounted on said housing and having a rim portion extending past the boundaries defined by said housing for supporting said housing on said surface;
   c. Second position wheel rotatably mounted on said housing with its axis of rotation oriented perpendicular to the axis of said first wheel, said second position wheel having a rim portion extending past said housing for supporting said housing on said surface;
   d. Transducer means connected to each of said first and second wheels, for generating digital position indicating signals indicating the degree of rotation of said wheels; and
   e. Flexible conductor means for connecting said transducer means to said computer, for conducting said position indicating signals to said computer while enabling unrestrained movement of said housing relative to said computer.

Reading patent claims may require knowledge of the relevant area of technology.

Why is claim drafting so tricky? The inventor cannot simply describe what the invention does. The inventor of the chair could not have patented it as simply "something to sit on." Rather, he has to distinctly claim it by listing the elements of the product or process. That is difficult. Try it with the chair.[5]

Questions

See if you can identify the inventions in these patent claims:

1. Is this a real invention or a fictitious patent claim? Hint: they say not to reinvent it.

   I claim an article of manufacture, comprising a disc that defines a continuous or substantially continuous circular outer periphery.

2. A tough one to improve on:

   I claim is:—

   1. In a bait trap the combination of a plain slab of wood a compound spring trapping frame a hinged catch lever and a bait hook staple substantially as herein set forth and illustrated by the accompanying drawing.

3. P-A-T-E-N-T.

   We claim:

   1. A transgenic non-human mammal all of whose germ cells and somatic cells contain a recombinant activated oncogene sequence introduced into said mammal, or an ancestor of said mammal, at an embryonic stage.

4. *Silly question*: How about this one? (Fellow non–chemical engineers, start at the end.)

   What I claim as new and desire to secure by Letters Patent of the United States is:

   1. The process which comprises (1) heating a mixture comprising (a) liquid polymeric dimethylsiloxane obtained by hydrolyzing a substantially pure dimethyl silicon compound containing two hydrolyzable groups and (b) from 5 to 25 per cent, by weight, based on the weight of the polymeric dimethylsiloxane, of a compound of boron selected from the class consisting of pyroboric acid, boric anhydride, boric acid, borax, and hydrolyzed esters of boric acid, the said heating being continued until a solid, elastic product is obtained, (2) adding a finely divided inorganic filler to the solid elastic product and 12 per cent, by weight, zinc hydroxide, based on the weight of the solid polymeric dimethylsiloxane, and (3) kneading the composition of (2) until a putty-like, elastic, plastic product is obtained.

5. And this one? Hint: the inventor was Rudolph Diesel.

> **What I claim as new is—**
> **1.** In an internal-combustion engine, the combination of a cylinder and piston constructed and arranged to compress air to a degree producing a temperature above the igniting-point of the fuel, a supply for compressed air or gas; a fuel-supply; a distributing-valve for fuel, a passage from the air-supply to the cylinder in communication with the fuel-distributing valve, an inlet to the cylinder in communication with the air-supply and with the fuel-valve, and a cut-off, substantially as described.

6. One more, if you are not dazed and confused by now. Hint: the inventor was Ferdinand Graf Zeppelin.

> Having now described particularly the nature of my invention, what I claim, and desire to secure by Letters Patent, is—
> **1.** In a balloon, the combination of a framework divided into separate compartments, with a main gas-bag in each compartment, adapted to expand and fill the same when permitted, and auxiliary gas-bags in the compartments for maneuvering, to permit the main gas-bags to retain their full quantity of gas unaffected by the admission of air, substantially as set forth.

Answers

1. The wheel. The drafter continued with some dependent claims, including

   Claim 4: The article of manufacture of claim 3, wherein the central column extends outward from each of the opposed faces a sufficient distance so that a human may rest a foot on respective sides of the disc on top of the central column (the B.C. wheel).[6]

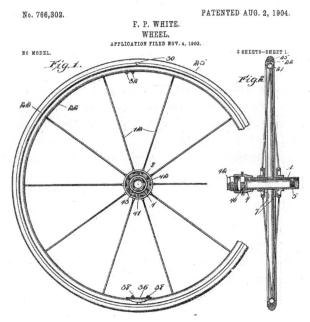

No. 766,302.

PATENTED AUG. 2, 1904.

F. P. WHITE.

WHEEL.

APPLICATION FILED NOV. 4, 1903.

NO MODEL.

3 SHEETS—SHEET 1.

Not the Wheel, Just a Reinvention

2. The mouse trap, of course.

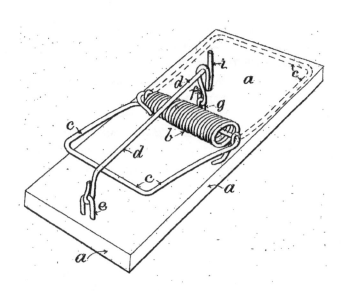

3. The Harvard Mouse (genetically engineered rodent—although note that the claim is broad enough to cover other mammals).
4. Silly Putty.
5. The diesel engine (no spark plug; rather, it squeezes fuel and air so much they ignite themselves).
6. Not heavy metal: the dirigible, also known as a zeppelin.

Description of invention
(show what you've got)

The applicant must describe the invention, disclosing how to make and use it. That is ironic because the point of a patent is that others are not allowed to make and use the invention without the inventor's permission, at least not until the patent expires. In pre-patent times, inventors often protected against copying by secrecy. Empirics and apothecaries kept their medicines secret, which was bad for the public (to the extent the potions really worked).[7] Some might disclose their remedies' ingredients, relying on their reputation as the best source to secure their success over copycats.

Peter Chamberlen invented forceps to aid in childbirth. That invention saved hundreds of lives during the 17th century. It could have saved many thousands. Over three generations, the Chamberlen family kept the use of forceps a trade secret, concealing the instrument and even blindfolding their patients at times.[8]

For many inventors, disclosure may be the biggest advantage from obtaining a patent. The vast majority of patents have no direct commercial value, viewed solely in terms of securing market power. Some simply represent unlucky bets, where applicants filed in the hopes that the invention would be commercially successful. But patents have many other types of value beyond securing a market. Patents are displayed to investors (such as showing venture capitalists that the start-up indeed has developed something), to customers ("our patented technology"), to competitors (as part of patent portfolios, which industry rivals use like missile silos). For many firms, far from being secrets to protect, technology is like having antlers to flaunt.[9] Inventors also seek patents not just to impress commercial parties but as a means to tell their story to scientific peers, and even historians. Disclosure can be the price the patentee pays, or the reward the patentee gets.

Because they disclose inventions, patent records can be a rich resource for researchers of all stripes. Economists have long used patent records in studying the relationship between technology and economic development. As *The Cambridge Economic History of Modern Britain* puts it, "One of the few available quantitative output indicators for technology" is the records of the patent office.[10] Patents also provide a source of information for the history of technology itself, as well as a useful source of technical information. Thomas P. Jones, an influential figure in early U.S. patent practice, "envisioned the Patent Office as a great repository of technical wisdom. He saw it on one hand as a museum in which the mechanic could trace the historical progress of the art and, on the other hand, as a collection which described the present state of the art."[11]

Patent records have been used to rethink the role of marginalized groups, as in *Mothers and Daughters of Invention: Notes for a Revised History of Technology*[12] and *A Hammer in Their Hands: A Documentary History of Technology and the African-American Experience*.[13] Patent records have facilitated specialized technological histories such as *Cotton: Origin, History, Technology and Production*[14] and *Glass: The Miracle Maker: Its History, Technology and Applications*.[15] Often the only remaining documentary evidence of an invention is its patent record.[16] Patents have played a role in forensic research as well. Art conservation scientists used patents on paints and pigments to conclude that certain paintings attributed to Jackson Pollock were actually painted after his death.[17] Patents even play a role in biographical research. The patents of Abraham Lincoln ("A Device for Buoying Vessels Over Shoals"[18]) and Albert Einstein ("Refrigerator") show less known sides of their personalities.

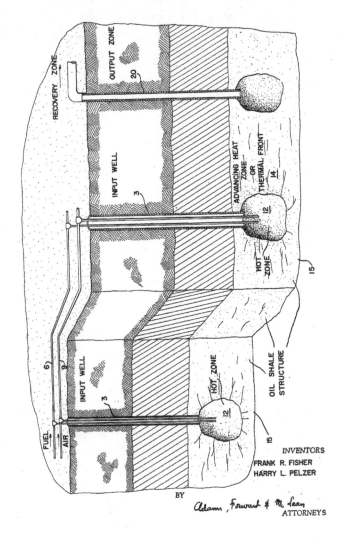

Feb. 5, 1957　　F. R. FISHER ET AL　　2,780,449

THERMAL PROCESS FOR IN-SITU DECOMPOSITION OF OIL SHALE

Filed Dec. 26, 1952

Fracking Patent

Patents may disclose information useful to competitors. In fast-moving areas of technology, such as biotechnology and computer science, companies may monitor patents and applications in their area of technology. That shows both what competitors have patent protection on and also what competitors deem important.

Patents may also disclose information useful for matters of public concern. Patents on fracking did both, revealing drilling techniques to competitors in the oil business and also making public information about drilling methods using chemicals and water at high pressure that raised environmental concerns.[19] However, the patent holder may limit disclosure of information

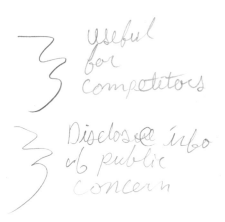

by people licensing the technology, which may undercut the public benefits, as scholars have pointed out.[20] If licensees have signed nondisclosure agreements (as to both the patented technology and their experience using it), while the patent sits silently in the files of the USPTO, the public may never hear important information.

The inventor must show that the invention is ready to be made. Otherwise, one could patent a mere idea. William Walcot received a patent in early times for a vaguely described machine to remove salt from sea water, which was more hope than reality. The vaporware invention itself evaporated. A patent issued on the basis of good intentions could deter other inventors from actually developing the invention.

```
<210> SEQ ID NO 19
<211> LENGTH: 20
<212> TYPE: DNA
<213> ORGANISM: Artificial Sequence
<220> FEATURE:
<223> OTHER INFORMATION: PCR oligonucleotides

<400> SEQUENCE: 19

atcggacgtg gacgtgcaga                                              20
```

What is claimed is:

1. An isolated recombinant anti-TNF-α antibody or antigen-binding fragment thereof, said antibody comprising a human constant region, wherein said antibody or antigen binding fragment (i) competitively inhibits binding of A2 (ATCC Accession No. PTA-7045) to human TNF-α, and (ii) binds to a neutralizing epitope of human TNF-α in vivo with an affinity of at least 1×10^8 liter/mole, measured as an association constant (Ka), as determined by Scatchard analysis.

of A2 (ATCC Accession No. PTA-7045), and (ii) binds to a neutralizing epitope of human TNF-α in vivo with an affinity of at least 1×10^8 liter/mole, measured as an association constant (Ka), as determined by Scatchard analysis.

8. A composition comprising the antibody or antigen-binding fragment of claim 1, and a pharmaceutically acceptable carrier.

9. The antibody or antigen-binding fragment of claim 1, which has specificity for a neutralizing epitope of human TNF-α.

Mouse Antibody Patent

The disclosure limits the claims. The applicant may amend the claims during the process of patent prosecution. That way, if the USPTO denies the application because the claims are too broad or otherwise defective, the applicant may amend the claims in order to receive a patent. Otherwise, deserving inventors would be denied patents simply because they drafted claims poorly. But an applicant may try to amend claims in order to cover the processes or products of others that became public after the application was filed.

Centocor filed a patent application on a pharmaceutical, a mouse antibody targeted at treating arthritis. Abbott, a competitor, later marketed a pharmaceutical for treating arthritis based on a similar human antibody. Centocor amended its patent claims to cover human antibodies, obtained a patent, and sued for infringement. A jury awarded $1.67 billion. The appeals court overturned the award. Centocor's amended claims were too broad because the application had only described a mouse antibody.[21]

The description must enable others to make and use the invention. The applicant need not have actually made the invention. Buckminster Fuller could patent his Undersea Island and Suspension Building without making them.

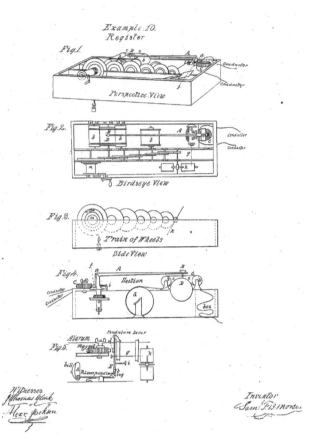

Internet, 1840?

Samuel Morse Did Not Invent the Internet

The patent claims are limited to the invention described. Samuel Morse's 1840 patent application supported his claims for various telegraph devices and for Morse code. But he threw in a final claim that went well beyond the invention he disclosed, claiming any means of using electromagnetism to communicate at a distance. That would have covered not only his invention but everything in telecommunications, through the iPhone and Internet and *Star Trek* communicators. That claim was invalid on the basis of being too broad.

Question

Buckminster Fuller never made an Undersea Island, but he patented it. How can he show he has the invention and show others how to make it if he never made it himself?

Answer

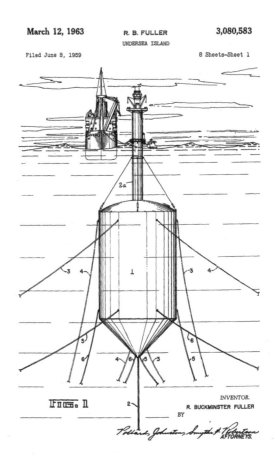

March 12, 1963   R. B. FULLER   3,080,583

UNDERSEA ISLAND

Filed June 8, 1959   8 Sheets—Sheet 1

FIG. 1

INVENTOR.
R. BUCKMINSTER FULLER
BY
ATTORNEYS.

The inventor must describe the invention and enable others to make it. If Fuller's disclosure did that, he need not also make the invention to prove his point.

Drawings of the invention

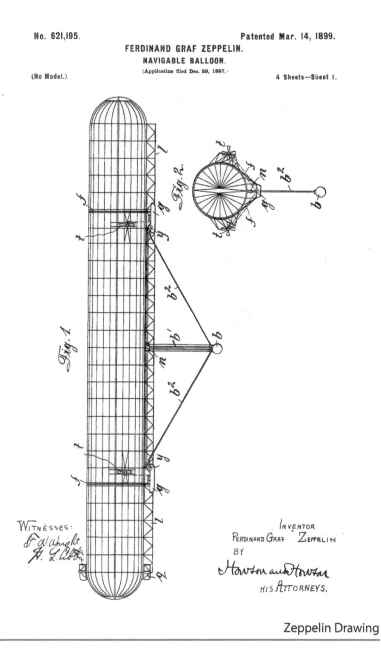

Zeppelin Drawing

Zeppelin's claim, as shown here, was not easy to read, but the drawings in the patent application helped.

Most applicants include drawings. Buckminster Fuller's patent on the geodesic dome describes the structure in great detail, and claims it concisely: "I claim a building framework of generally spherical form in which the main structural elements are interconnected in a geodesic pattern of approximately great circle arcs intersecting to form a three-way grid defining substantially equilateral triangles." Huh? The drawings are worth many words.

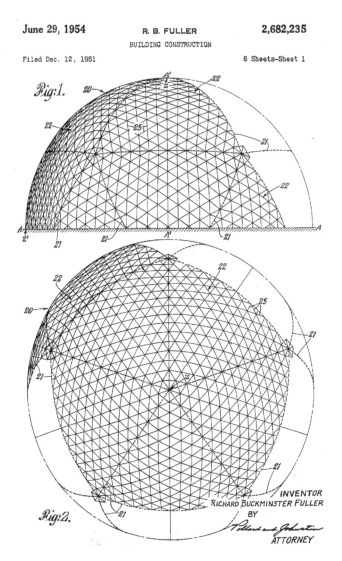

Early Buckyball?

During the 1800s, inventors were required to include models with their applications. Models are no longer (usually) required. Where would the USPTO put them all? Sometimes, however, the USPTO will require a model. If the applicant claims a time machine or perpetual motion device, the USPTO may require submission of a working model. So far none have been submitted. Maybe a time machine will be sent back from the future one day.

In some areas, to "describe" the invention, applicants submit samples of the invention. Biotech researchers may develop a new microbe or chemical and explain how it has a useful effect, yet not be able to specify just what they have in terms of a chemical formula or other literally written description. By depositing a sample, the inventor can show both possession of the invention and enable others to make it. The sample goes in an authorized repository, not in the desk drawer of the examiner.[22]

To Patent or Not to Patent

Since new developments are the products of a creative mind, we must therefore stimulate and encourage that type of mind in every way possible.

George Washington Carver

Why patent?

Why do people buy land? There are many reasons: to invest, to build on, to farm, for conservation, to keep up with the Joneses, to keep away from the Joneses, to mine, to have a home, to reduce taxes. . . . Patents likewise have many uses, commercial and otherwise.

Some patents create a monopoly: a product or process with no ready substitutes, so the patent owner can charge a monopoly price. Many drugs fall into this category, because the drug manufacturer has two protections: the patent (so no one else can sell the same drug) and Food and Drug Administration approval (it takes years and millions of dollars to get a drug approved, so possible competitors face high market-entry costs). Only a small percentage of patents, however, have true monopoly value.[1]

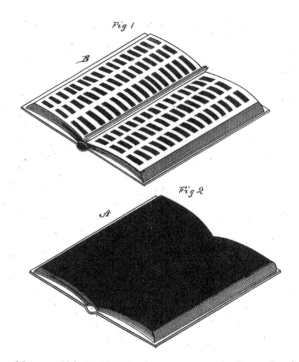

Twain's Ill-Fated Patent, U.S. #140,245—Improvement in Scrap-Books

Most patents do not secure a market. Mark Twain went bankrupt trying to make a fortune with his patented scrapbook. Einstein patented a refrigerator and a camera; they did not have the success of his theoretical physics.

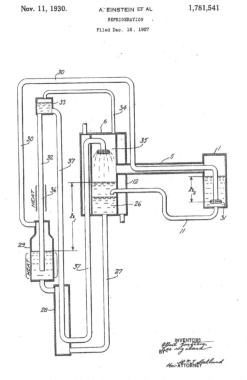

Einstein's Refrigerator—Not Cold Enough for Einstein-Bose Condensate

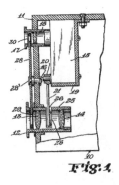

Fig.1

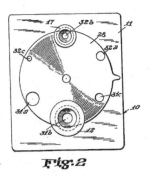

Fig.2

The "et al" after "G. Bucky" Is Albert Einstein—He Got Second Billing to Gustav Bucky (Who Was Not the Inventor of Buckyballs)

On the flip side, a patent is by no means necessary for commercial success. Blockbuster products in every field have succeeded without patent protection: iconic toys like the Easy-Bake Oven, Hula Hoop, and Radio Flyer; office products like Wite-Out; software like the early versions of Microsoft Windows and Facebook (both have lately become avid in getting patents).

Even a valuable invention does not market itself. Thomas Edison's commercial success came as much from his marketing and organizational acumen as from the inventions themselves. To sell the light bulb, for example, required getting people to adjust to new things. As a plaque used in hotel rooms read,

> This room is equipped with Edison Electric Light. Do not attempt to light with match. Simply turn key on wall by the door.

Edison's tactics in dealing with competitors are storied. He promoted the use of the name of his competitor "Westinghouse" as a synonym for "electrocute." Edison pushed through infrastructure to support the use of his inventions, such as starting power companies so that the necessary electricity would be there for light-bulb buyers. By contrast, Thomas Davenport's 1837 electric motor was a technological breakthrough, but it made little money because there was, as yet, no electric grid or even practicable batteries.

T. DAVENPORT.

Electric Motor.

No. 132.   Patented Feb. 25, 1837.

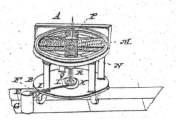

Ahead of Its Time

Dec. 11, 1973   D. C. BOYD   **3,778,335**
SODIUM ALUMINOSILICATE GLASS ARTICLE STRENGTHENED BY
A SURFACE COMPRESSIVE STRESS LAYER
Filed Sept. 2, 1971   2 Sheets-Sheet 1

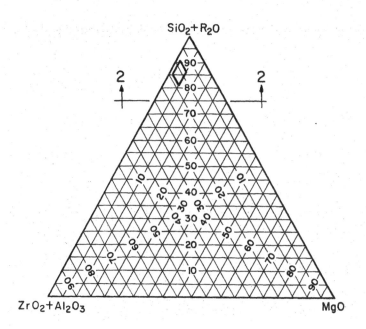

Sodium Aluminosilicate (Scientific Name = Gorilla Glass)

Gorilla Glass was also ahead of its time. Corning Glass developed an incredibly strong glass, but the applications were few and expensive. Unbreakable windshields for cars sound good, but it is actually better for the windshield to give way to a human head. Decades later, however, superstrong glass was perfect for smartphone screens. Fortunately for Corning, adapting the glass required Corning's expertise and later patents.[2]

Most patents do not grant a true monopoly: a single seller of a product or service with no ready substitute. A monopolist has no competitors, so it can charge a premium. An inventor may patent a new mousetrap, but there are plenty of other types of mousetraps to compete with. But a patent may protect some features of a product, which in turn may help it compete against other products (which may have their own attractive features, some patented). Or the patent holder may simply license the patent to parties that make products or processes covered by the patent. A typical computer manufacturer licenses hundreds of patents from various patent holders.

A patent portfolio (a large number of patents in the relevant area of technology) may help a company in many ways. Companies like Facebook and Microsoft and Apple paid little attention to patents in their early stages. All have loaded up with thousands of patents in recent years, both by patenting their own inventions and buying up patents, often from companies that had technical success followed by financial problems, such as Kodak, Motorola, and AOL. A patent portfolio can

- *Bring in licensing revenue*: IBM paved the way with licensing agreements for many of its patents, and other technology companies are following suit.

- *Provide defense*: A company with a bristling patent portfolio can use those patents to menace others that seek licensing fees for their patents.
- *Allow cross-licensing*: Patent holders cross-license each other's patents, not unlike nonaggression pacts between overarmed nations— with competition-dampening effects on the companies in the market who are left out.
- *Act as a signal*: Just as a peacock's magnificent tail signals that the animal is robust, so a potent portfolio attracts investors and warns potential competitors. Sellers often use patents to signal to potential buyers that the product represents a big investment, as in, "our patented product."

Some companies own nothing but patents. Intellectual Ventures holds thousands of patents, both developed in-house and bought, and seeks licensing revenue in many industries. It has less to fear from the patents of others for the reason that it does not make or sell anything, although there are patents on methods of licensing too(!), and anyone who uses computer hardware and software is infringing many patents. Are such "nonpracticing entities" a good thing? Some label them "patent trolls," contributing nothing and simply taxing the productive, like the evil troll living under the bridge. Others extol them as sharpening the incentive to innovate, which is what drives the modern economy.

The biggest factor influencing whether a company seeks patent protection may be peer pressure.[3] Companies tend to behave like schools of fish. In industries where patenting is common, companies tend to seek patents. Over time, the same industry can move from patent ignoring to patent seeking. In software, companies like Microsoft went for decades without seeking many patents, as did its competitors. Then software companies became aware of the possible risks and rewards of patents and now are quite active in seeking patents. They tend both to patent their own work and to seek to buy the patents of others (such as competitors who have foundered). To take another industry, Swiss watchmakers relied on trade secrets rather than patents. Swiss law was not friendly to patents, and Swiss watchmakers found that their English competitors were not able to reverse-engineer their work. For that reason, the progress in watchmaking is less documented in the patent records than are many other industries.

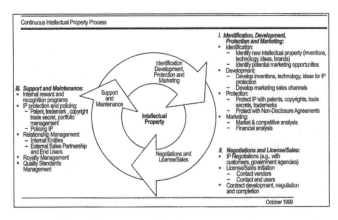

Method for Determining Marketability (Frank et al., "System and Method for Determining the Marketability of IP Assets," U.S. Patent 7,840,498. Presumably This Patented Invention Could Value Itself.)

The role of patents varies by industry: "In the pharmaceutical industry, the medical device field, or the traditional mechanical field, an individual may only have one or two patents covering his invention. In IT, however, one product regularly involves the combination of fifty, one hundred, even one thousand, or—as Intel lawyers themselves say with respect to their own core microprocessor—five thousand different patent rights."[4] In information technology industries especially, many factors combine to leverage the risks of infringement and the rewards of holding even uncertain patents. "Royalty-stacking" refers to the need for multiple patent licenses for some products, especially computer/phone technology. "Reach-through" clauses, more common in areas like biotech patent licenses, may give the patent holder rights over any technology the licensee develops.[5]

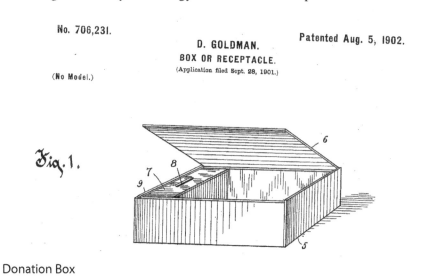

No. 706,231.

D. GOLDMAN.
BOX OR RECEPTACLE.
(Application filed Sept. 28, 1901.)

Patented Aug. 5, 1902.

(No Model.)

Fig. 1.

Donation Box

If a patent yields less funds than hoped, it may be donated to get a tax deduction. Yes, there's an app for that: "A system and method for facilitating the donation of intellectual property assets (e.g., patents)" (U.S. Patent Application 10/078,135). Perhaps that invention will be used to donate itself.

Patents also appear in marketing. Advertisements in every type of media tout patents obtained on the product. An ad for a hybrid car featured an inset citing its "Patent No. US6687593 Regenerative Braking."[6] The advertising agency did not really think the typical consumer wanted the patent number, which one would need only if one wished to look up the patent. However, citing the patent number lent the ad a flavor of authenticity and authority, the suggestion that the USPTO, after due examination, had attributed special qualities to this invention. It is a modest patent holder that simply puts the required patent notice on the product itself.[7] Advertisements refer conspicuously to "our patented technology," as with the hybrid automobile's regenerative braking system. One might think the car seller invented regenerative braking, but upon inspection of the patent cited, it becomes clear that it simply covers one type of regenerative braking.[8]

Businesses may flaunt patents as an indicator of how innovative they are. As one CEO-turned-presidential-candidate put it, "We doubled the size of the company from $44 billion to $88 billion. . . . We tripled the rate of innovation to 11 patents a day."[9] Start-up businesses often seek patents as a sign to potential investors that the start-up has a genuine innovation.

The power of patents may also subtly suggest other things to consumers. But the grant of a patent, strictly speaking, speaks to none of the qualities of the product.[10] A consumer is likely to be interested in whether a product is safe, efficient, better than other products on the market, pleasant to look at, and so on. Yet whether a product is patented depends on none of those qualities. An inventor is entitled to a patent on his wooden combination lock if it is new, useful, and nonobvious. If it has not been done before, is useful, and would not have been obvious to one working in that area of technology, he gets his patent. The lock may not work better than ones on the market, and need not even work well at all. It need not be more efficient than ones already available. It may be hazardous, even, or used for fraudulent purposes. But somehow the many invention stories we have heard suggest to us that a patented product must only be new. No one would buy a book because it is copyrighted (indeed, freedom from copyright protection technology is a definite selling point for music, games, and many other copyrighted works). But patents sell.

Beyond business

Beyond business, people seek patents for more personal reasons. Patents are often displayed like trophies. NASA patented much space technology, even at times when it had no competitors, other than the USSR, which cared little for U.S. patents. Author biographies in *Scientific American* list patents along with scientific papers. In academia, the old saying might be amended to "publish, patent, or perish" because professors list their patents on their CVs and home pages. Many a business titan has the odd patent, often for an invention with no likely commercial advantage, just to show another facet of her personality. Some people seek a patent simply because they are proud of their invention.

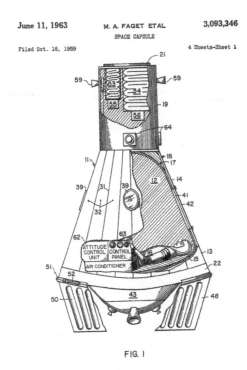

Patent Owned by NASA, U.S. Government

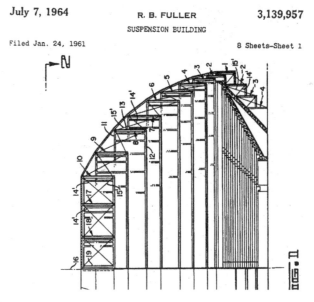

July 7, 1964   R. B. FULLER   3,139,957
SUSPENSION BUILDING
Filed Jan. 24, 1961   8 Sheets—Sheet 1

Bucky Building

Some have used patents to make a record of their work, even where commercial prospects are remote. Buckminster Fuller patented an Undersea Island and a Suspension Building (think of the Golden Gate Bridge but as a building). Fuller sought patents for such visionary inventions, not to prevent others from exploiting his work (neither Fuller nor anyone else built an Undersea Island or Suspension Building), but as a record of his work.

Bell Labs sought a patent on Claude Shannon's robotic mouse Theseus, which could navigate its way out of a maze. Bell Labs did not think the invention had commercial value, but diverted the resources from its more likely money getters in order to make a record of the innovation, one of the first devices that could learn with electric circuits.[11] At the other extreme, the United States sought to control future uses of atomic power by occupying the field with patents.[12]

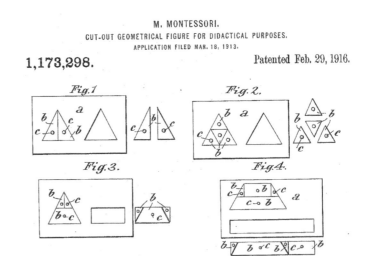

Patents with Promotional Value

Patents resemble merit badges in other areas. Patents receive increasing attention and weight, from advertising to academia, even as the many problems of patents have become better known.[13] A story in *Popular Mechanics* about an inventor mentions his "more than 100 patents."[14] *Scientific American* author biographies often give as much space to the patents awarded a scientist than to more traditional credentials such as publications in peer-reviewed journals. Perhaps it is more effective to list the patents a researcher has than to list the specialized journals she has published in. Patents have entered into tenure disputes, with the number of patents a professor received (and even the applications he had pending) cited as evidence that he should have been tenured.[15]

A strange tale found a scientific peer reviewer giving deference to patent examiners. A researcher submitted a paper to the *Physical Review Letters*, a leading peer-reviewed journal. The paper, "Optical Conformal Mapping," described a technique to guide electromagnetic waves around an object, thereby disguising the object. One of the reviewers recommended against publication, partly on the grounds that another team had reportedly "filed a patent" on similar work.[16] Peer reviewers should be leading experts in an area of science. For a peer reviewer to rely on a patent examiner's rumored opinion in assessing the importance of a paper is exactly backward.

Why not patent?

Coca-Cola might have patented its formula, perhaps in 1903, when it switched from cocaine to caffeine (in a sense, the original "New Coke"). That would have disclosed the formula to the world. After the patent expired around 1920, the world would have been free to copy Coke. But Coca-Cola did not patent its valuable formula. It keeps it as a trade secret—for over 100 years and going strong.

Many inventors have the choice between relying on trade secrets or patents.[17] Examples include software that gives an edge to a hedge fund, a manufacturing process to make electronics, and a process for making chemicals. A patent applicant pays thousands of dollars to disclose valuable information to the world but in return receives rights in the information for about 17 years. The inventor could simply exploit the information in secret. That would avoid patent fees and disclosure. As one technologist put it, "We have essentially no patents in SpaceX. Our primary long-term competition is in China. If we published patents, it would be farcical, because the Chinese would just use them as a recipe book."[18] But trade secrets have costs of their own: the costs of the security program, the restrictions on passing information around, the risk that others will figure the information out—not to mention the loss of reputation points.

Noncommercial inventors likewise prefer not to disclose their inventions to the public. If Q from the James Bond movies has a real-life counterpart, she would not file patent applications telling the world about her agents' secret weapons and how to make them.

Some inventors do not have the choice between patents and trade secrets. If the invention is a new toy, once sold it is no longer secret. Some information must be disclosed to be commercialized. The FDA will not allow a drug to be sold with secret ingredients.

Other inventors may pass on patents simply because of the cost. The basic filing fee is $380, then $620 for the examiner to search for prior art, and $250 for examination. If the patent is approved, the inventor must pay an issuance fee of $1,740 in order to receive it. Then, to keep the patent in effect, the patent owner must pay maintenance fees: $1,600 after three and a half years, $3,600 after seven and a half years, and $7,400 after eleven and a half years. For more complex applications, there are additional fees.

All the fees are reduced by 50 percent or more if the applicant is a small entity or even a "micro-entity." Microsoft is not a micro-entity. Rather, individual inventors, small businesses, and universities (Ohio State: a micro-entity?) all receive the reduced rate. Those are simply the USPTO fees. A wise applicant will have a patent lawyer or patent agent draft and prosecute the application, which can multiply the costs. An applicant may wish to have patent rights in other countries, which will further increase fees—for filing fees, representation, and translation from English (well, legalese mixed with tech talk) to the relevant language. Finally, simply having a patent may not bring in money. Enforcing the patent in court costs a lot (of course, defending a lawsuit is also costly, so this can be a factor that encourages infringers to agree to pay for a license).

It will take several years for the patent to issue. In fast-moving industries like software, patents may issue after the product is passé. Other factors, like being first to market or putting out the best product, may offer more advantages than patents.

An inventor may not think of patenting until it is too late. After selling a product with increasing success, the inventor might think of patenting it—but then public disclosure will have cost the right to patent. Other inventors may not think the invention is patentable. The electronic spreadsheet was once the "killer app" that spurred the sales of personal computers. Its inventors did not file for a patent because the courts had not yet made it clear that software was patentable.

Some inventors decline patents, believing that their invention should remain in the public domain. Jonas Salk, inventor of a polio vaccine, told Edward R. Murrow, "There is no patent. Could you patent the sun?" Tim Berners Lee, inventor of the hypertext protocol for the World Wide Web, chose free standards over patent rights (and thereby greatly speeded adoption of those standards). George Washington Carver, whose unpatented inventions transformed agriculture, reportedly said, "God gave them to me. How can I sell them to someone else?" Benjamin Franklin's many inventions included a revolutionary stove and the lightning rod. Franklin did not seek patents or challenge others who seemed to patent his work, "having no Desire of profiting by Patents myself, and hating Disputes."

Most computers now use von Neumann architecture, developed at the Institute for Advanced Study (Einstein's institution, but he stayed in the theory department). The machine's name, MANIAC, commented wryly on other early machines like the ENIAC and ILLIAC. The highly new and nonobvious useful device could predict the weather 24 hours in advance, but took 24 hours to do the calculation. It also could simulate nuclear reactions and so played a key role in the early Cold War with the USSR. There was discussion about whether to

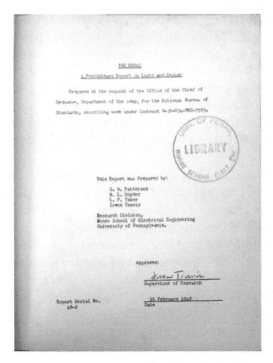

THE ENIAC

A Preliminary Report on Logic and Design

Prepared at the request of the Office of the Chief of
Ordnance, Department of the Army, for the National Bureau of
Standards, describing work under Contract W-36-034-ORD-7593.

This Report was Prepared by:

G. W. Patterson
R. L. Snyder
L. P. Tabor
Irven Travis

Research Division,
Moore School of Electrical Engineering
University of Pennsylvania.

Approved:

Irven Travis
Supervisor of Research

Report Serial No.
48-2

16 February 1948
Date

John von Neumann, First Draft of Report on EDVAC (1945)

patent the device, and about whether the IAS or the computer team would own the patents. Eventually, John von Neumann decided not just to forgo patent protection, but to deliver the key paper describing the machine's design to the USPTO, to ensure that others could freely use the design. The United States, although keeping a tight lid on atomic weaponry information, did not so limit the spread of information on computer technology.

bad-ass

Some inventors are just not that concerned with commercializing their work. As Nobel Prize–winning physicist Isaac Rabi put it, "I and my friends sat around and talked about what we'd do if we had a million dollars. I thought and thought and finally I said, 'I think I'd buy a new hat.'"[19]

Scientists have become more likely to patent.[20] The 1999 Nobel Prize winners in medicine and chemistry emphasized that their work decades earlier had been driven by a desire for knowledge, not profit, to explain why they had not patented their discoveries. But recently, researchers at MIT reported that they had inserted a gene into a transgenic mouse, improving its memory. Patents pervaded the story. The gene used was already covered by someone else's patent, and the researchers, long before they told the *New York Times* or the rest of the world about their work, had filed for another patent on this new use of the gene. Scientists and engineers are not patent *naifs*.

Graphene, a molecule-thick sheet of carbon, may have many commercial applications. But its discoverer settled for a Nobel Prize and decided against patenting. As he put it to the science journal *Nature*,

We considered patenting; we prepared a patent and it was nearly filed. Then I had an interaction with a big, multinational electronics company. I approached a guy at a conference and said, "We've got this patent coming up, would you be interested in sponsoring it over the years?" It's quite expensive to keep a patent alive for 20 years. The guy told me, "We are looking at graphene, and it might have a future in the long term. If after ten years we find it's really as good as it promises, we will put a hundred patent lawyers on it to write a hundred patents a day, and you will spend the rest of your life, and the gross domestic product of your little island, suing us." That's a direct quote.

I considered this arrogant comment, and I realized how useful it was. There was no point in patenting graphene at that stage. You need to be specific: you need to have a specific application and an industrial partner. Unfortunately, in many countries, including this one, people think that applying for a patent is an achievement. In my case it would have been a waste of taxpayers' money.[21]

Patents are just one of many spurs to inventors. Some, like Moties (of *The Mote in God's Eye*, the science fiction novel), instinctively tinker to improve whatever crosses their path. Scientists and engineers innovate as their work. Inventors may make life better for others while gaining the esteem of their peers and satisfaction of internal drives. Most businesses sell unpatented goods and services.

Some prizes go to achievements in a defined area, such as the various Nobel Prizes. Others target specific technology. The X Prize went to the first nongovernmental team to put people in space. The Bill & Melinda Gates Foundation sponsored a competition to "Reinvent the Toilet," to help the millions of people worldwide without sanitary facilities, a public health issue. The winning design (from CalTech) uses solar power to recycle water and turn waste into fertilizer and fuel. That could improve millions of lives, with nary a patent in sight.

Patent Litigation: The Sport of Kings[1]

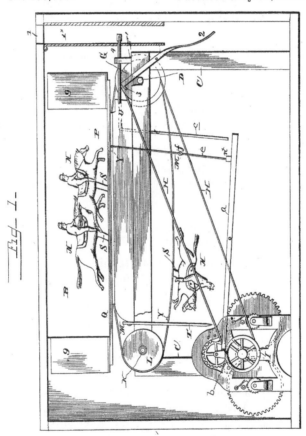

(No Model.)
3 Sheets—Sheet 1.

E. J. LUMLEY.
TOY HORSE RACE MACHINE.

No. 433,012.
Patented July 29, 1890.

Smaller Scale Sport of Kings

After submitting the detailed application, spending a ton on fees, and negotiating with the patent examiner over several years, the inventor finally receives his patent. For many inventors, the patent, an impressive document complete with a red ribbon, sits on the wall and is listed on the résumé, and that's it. Most patented inventions are not commercialized. Most patents do not bring in licensing fees. There is no need to enforce most patents because no one is infringing them or wants to make the patented invention.

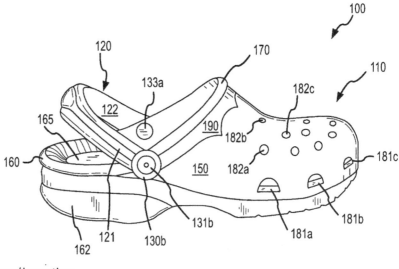

Crocs' Invention

Many inventions do have value. Suppose Crocs had received its patent on "breathable footwear pieces," which claims the invention as follows:

I claim a footwear piece comprising

A base section including an upper and a sole formed from moldable foam; and
A strap section formed of a molded foam material attached thereto.[2]

In the meantime, many similar sandals have appeared on the market. What rights does the patent give Crocs? Anyone who makes, uses, sells, or imports the invention in the United States infringes the patent.

Crocs's patent is infringed by:

- a manufacturer making the sandals
- kids wearing the sandals(!)
- wholesalers and retailers selling the sandals
- foreign manufacturers or sellers, if they import the sandals into the United States

Crocs won't sue the kids. Crocs would win no friends and only teeny money suing consumers. But if Crocs thinks someone is infringing, it is up to Crocs to take action. No patent enforcement bureau monitors possible patent infringement. Rather, Crocs can sue infringers—in court or, for importers, before the International Trade Commission.

History is littered with inventors who were unsuccessful in enforcing their patents. Charles Goodyear's process for vulcanizing rubber was so elegant that it was easily copied.[3] He had little luck enforcing his patents. The vastly successful Goodyear Tire & Rubber Company? Founded decades later by others and simply named after Goodyear. Eli Whitney likewise had little success collecting from the many who infringed his patent on the cotton gin, despite the influence it had on agriculture and U.S. history.

Filing a lawsuit is not the best first step. Patent litigation is expensive, especially if expert witnesses are required due to complex technology or tricky marketing analysis. Crocs's first step may simply be to send a letter seeking either that the infringer stop selling the sandals or offering to license patent rights. Sometimes that works. The letter recipient likewise does not want to get into expensive litigation and may acknowledge Crocs's rights. They may cease infringing or negotiate with Crocs to pay for a license agreement. Or they may ignore the letter.

There is a hazard in asserting patent rights. If Crocs sends a letter alleging that Footies is infringing its patent, Crocs had better be ready to go to court. Once there is a dispute, Footies may file a lawsuit for a declaratory judgment that Footies is not infringing. You might ask why Footies would want to drag itself into court. One reason would be if Footies wants to clarify its rights. Another reason might be to control where the litigation takes place, by filing in the preferred judicial district. Or Footies may simply think that going to court will give it leverage to get Crocs to settle on favorable terms.

Suppose Crocs does sue a sandal seller—to get the seller to shut down, to collect money, and to encourage others to respect Crocs's patent rights. In that litigation, the court is likely to decide several issues:

- **Claim interpretation:** what inventions the patent claims cover
- **Validity:** whether the patent is valid (i.e., whether the patent application properly described and claimed a new, useful, nonobvious invention)
- **Remedies:** if there is infringement, what money and other remedies would apply

Claim interpretation[4]

> *Vizzini*: Inconceivable!
> *Inigo Montoya*: Somehow, I don't think that word means what you think it means.
>
> William Goldman, *The Princess Bride*
> (the book plus the movie adaptation)

> 'Cause you know sometimes words have two meanings.
>
> Led Zeppelin, "Stairway to Heaven"

Crocs's patent rights apply only to the invention claimed in its patent. The other sandal seller may argue that its sandals are not covered by Crocs's patent claim. Perhaps it argues that its material is not "foam" but rather a type of plastic. Or that its base section does not have an upper and a sole (as in Crocs's claim) but instead a new type of base. Or its strap can be easily unhooked and so is not "attached thereto."

Claim interpretation is an issue in most patent cases. The court cannot simply rely on past cases. Every patent claim is unique, because otherwise the claimed invention would not be new. Courts often have to read patent claims and determine how those words apply. All law involves words, but every patent depends on the meaning of the words in its particular claims. Because each applicant may draft his own claims, and even define the meanings of the words, interpretation is needed in every single case.

In some areas of law, documents and terminology are more standardized. High-tech law has not moved far from the 18th century, when there was no fine distinction between lawyers and writers. Samuel Johnson, the author of the best known early dictionary and a leading 18th-century literary figure, was also a legal thinker.[5]

Some examples of claim interpretation issues:

- In a vehicle navigation system, whether "coordinates" meant only position defined by longitude and latitude, or could be defined by distance from a point.[6]

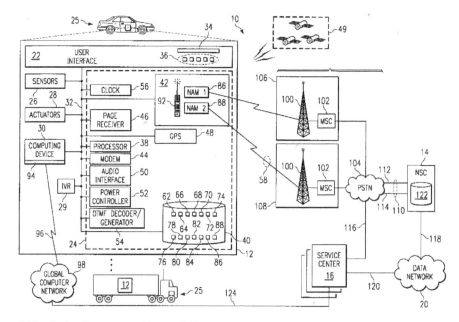

"What's the Frequency, Kenneth?"

- In a patent on a biopsy device, whether a part was "detachable" if it could be taken off, or only if it could be taken off without destroying the part.[7]

Fig. 11

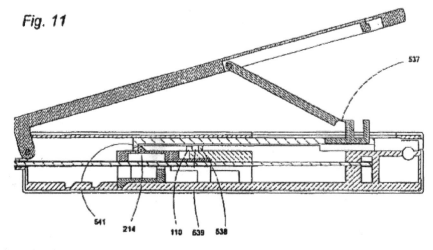

Detachable?

- In a circuit board, how far apart a reducer could be from a load and still be "near" it.[8]

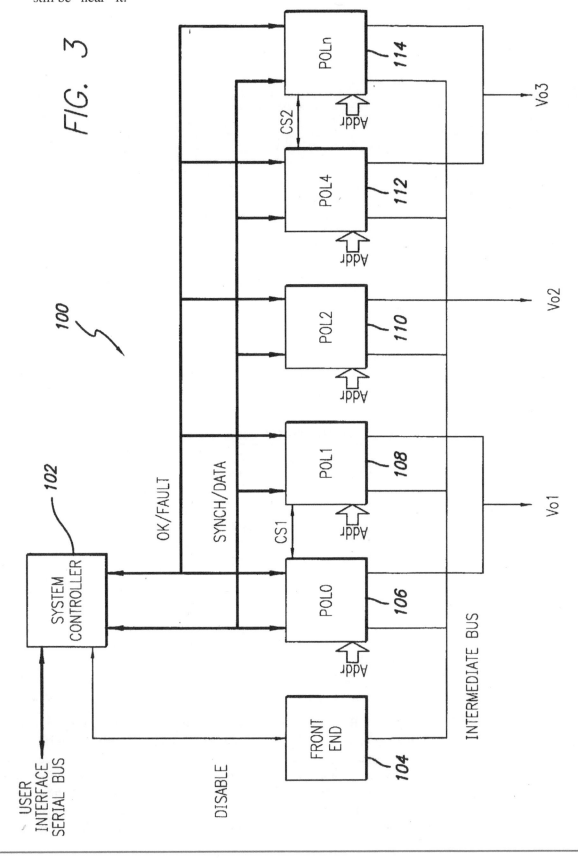

FIG. 3

Near?

- The distinction between "malleable" and "resilient" for wires in medical grafts.[9]

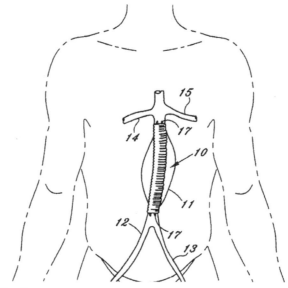

Malleable Wires?

- Whether the term "animal" includes "humans."[10] (*Hamlet*: "What is a man, If his chief good and market of his time, Be but to sleep and feed? A beast, no more.")

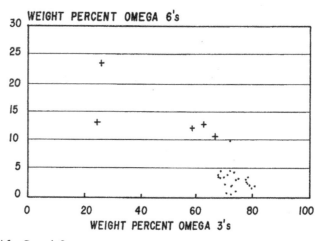

Patented for People?

- A human is an animal—so held the patent appeals court.[11]

Martek Biosciences held a patent on "methods for increasing the concentration of omega-3 HUFA in animals by feeding them microorganisms of the order Thraustochytriales." Martek had patented a process of increasing the healthy omega-3 fats in animals by feeding them enriched algae. Nutrinova used similar methods on nutritional supplements for people. Yummy, algae. When sued for infringement, Nutrinova argued that its supplements fell outside the patent claim because Martek's patents covered feeding animals, and humans were not animals. The court held that, as used in the patent, the term "animals" included humans.[12]

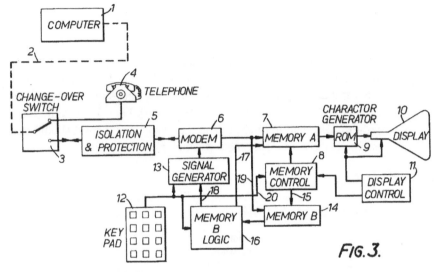

FIG. 3.

World Wide Web, 1977?

Question

British Telecom (BT) held a patent (applied 1977, issued 1989) on an Information Handling System and Terminal Apparatus Therefor. Highlights of the relevant claim:

What is claimed is

> A digital information storage, retrieval and display system comprising:
>> A central computer in which plural blocks of information are stored
>> Plural remote terminals
>> Further memory providing digital data indicative of complete addresses of blocks of information on the central computer.

BT read the claim to cover hyperlinks, as used on the World Wide Web. BT first sent cease-and-desist letters to Prodigy and other Internet service providers. After a cool reception, BT sued for infringement. Who won? Hint: WWW does not stand for "British Telecom."

Answer

BT lost.[13] The court found several reasons why the claim did not cover the Internet. The Internet did not have a "central computer," and the claim could not be read as referring to lots of central computers linked together—which would sort of defeat the meaning of "central." A computer that can run all manner of software, including surfing the Web, is more than a "terminal" (a term more often used for relatively dumb devices). The information on the Internet is not arranged into segregated "blocks of information." Rather, it is intermingled on each page. URLs are not complete addresses (as claimed in BT's patent), because a URL is a virtual address, which must be sent to a domain name server to get the IP address to which it refers.

The BT case shows how claims, especially software and other methods-type claims, can be read broadly by hopeful patent owners. Suppose the deed to some farmland read "the tract between Spring Creek and Gorgeous Gorge," a distance of about 1 kilometer. The owner might argue that "between" means all the land between them, going the other way all around the Earth (40,008—1 km). The owner might think it can literally be read that way—but a court should not. And the same is true for patent examiners.

Validity of the patent

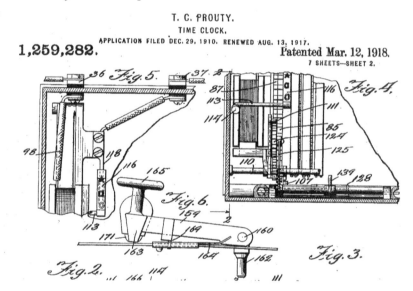

The Patent Examiner Has Less Time and Fewer Resources to Put into Examining Patents than a Defendant in a Lawsuit Does

If sued for patent infringement, the defendant may argue that the patent is invalid. Even though the USPTO has issued a patent, it may still be challenged in court. The court can revisit all the requirements of patentability: whether the invention was new, useful, and nonobvious, and whether the application properly disclosed how to make and use the invention.

The court also revisits the job done by the patent examiner—but the court has many more resources available than the examiner had. Patent examiners may have a few dozen hours to examine a patent application, and they will consult only whatever resources are at hand. When a patent holder seeks millions of dollars for alleged patent infringement, the other party has good reason to invest thousands of hours to scour the globe and pick the brains of experts to try to find publications or public uses that show that the invention was not new or was obvious. That is why some call patent litigation "The Sport of Kings."

New
Useful
nonobvious
How inventiones
to be used,
purpose

No. 707,982.
H. H. TAYLOR.
HEAD OR SEARCH LIGHT.
(Application filed June 2, 1902.)

Patented Aug. 26, 1902.

(No Model.)

2 Sheets—Sheet 1.

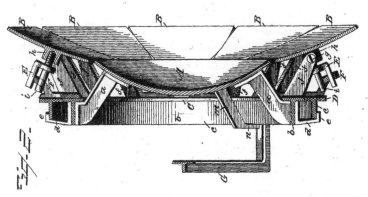

Someone Sued for Patent Infringement May Search for
"Prior Art" That Shows the Patent Is Invalid

In about one case in three, the court rules the patent invalid. That raises this question: There are about 2 million patents in force; are one-third of them invalid? That is possible. The patent examiner has little time and few resources to find prior art and to think of other reasons why an applicant may not be entitled to a patent. It makes sense for the USPTO not to spend millions of dollars of resources examining each patent perfectly: it constitutes what might be called "rational ignorance."[14] A defendant in court, by contrast, has the incentive to put considerably more effort into it. It may also be that the patents that get litigated are more questionable than average. Defendants faced with clear infringement of valid patents may be more likely to settle before the case gets decided by the courts.

Question

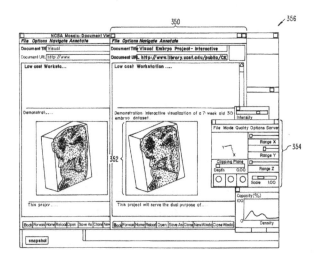

U.S. Patent 5,838,906: Eolas Browser Patent

Eolas holds a patent on the interactive Web browser. When Eolas sues MySpace for infringement, MySpace shows others had similar software before Eolas. Eolas can show it knew nothing of the other browsers. Would that help Eolas?

Answer

No. An invention is not patentable if it was not new or would have been obvious based on technology in the field, and this rule stands whether or not Eolas knew of the technology.

Patent infringement does not require copying

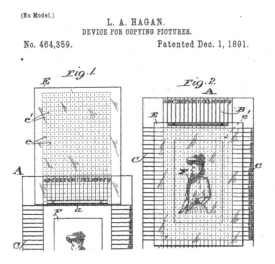

Patent Infringement Does Not Require Copying

Unlike copyright, patent infringement does not require copying.[15] Many patent infringers develop the technology independently, knowing nothing of the patent. This rule echoes the novelty requirement. An inventor is likewise not entitled to patent an invention that is already public even if the inventor did not know of the invention and reinvented it.

Question

Little Fish sues Microsoft for infringement of a file management patent. Would it absolve Microsoft if it could prove that it developed its file management technology itself without any knowledge or use of Little Fish's technology?

Answer

No. Patent infringement does not require copying. With software patents especially, this can be tricky. There are many thousands of software patents, which can apply in many areas.[16]

Do you think a second inventor should have to pay the first inventor for patent infringement even if she made the invention independently?

Remedies for infringement: Money and orders to stop

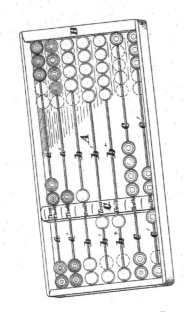

H. FITCH.
Caloulators.

No. 232,482.

Patented Sept. 21, 1880.

For the Jury to Add Up Money

An infringer may be ordered to pay money. The amount payable may be reckoned by the profits made from the infringement or by a licensing fee. Blatant infringers may have to pay an enhanced award, up to triple the actual losses, plus attorney's fees.

No. 648,621.

J. M. HOOPER.
STRAIT JACKET.
(Application filed July 24, 1899.)

Patented May 1, 1900.

(No Model.)

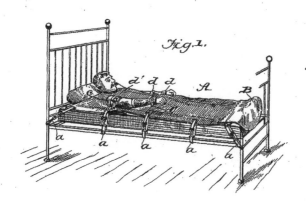

The Court May Order the Infringer to Cease and Desist

If an award of money does not protect the patent holder, the court may also order a stop to the infringing activity.

Question

If Microsoft's Windows 8 infringes Little Fish's patent, how do we figure out how much Microsoft will pay? Can Little Fish get an order that Microsoft must stop distributing Windows 8 until it reaches a deal with Little Fish or redesigns Windows 8 in a way that does not infringe?

Answer

Little Fish is entitled to a royalty. The court will try to figure out what the parties would have agreed on had Little Fish licensed its patent to Microsoft.

The court will not order Microsoft to stop selling a product in a case like this where only one part of a complex system infringes. Rather, a royalty should compensate Little Fish.

Trademarks

Trademark Infringement

Not Trademark Infringement

Cause I'm Slim Shady®,
Yes I'm the real Shady®,
All you other Slim Shadys are just imitating.[1]

Marshall Mathers (a.k.a. Eminem)

As a quick example of trademark rights, consider that Mars, Inc., owns the trademark M&M for candy. W&W is not the same, but used on a similar sized package of candy, it infringes the M&M mark because it is likely to cause confusion or deception.

Marshall Mathers adopted the name Eminem for his rap rhapsodizing. "Eminem" sounds exactly the same as "M&M." That's why **M**arshall **M**athers chose it. But EMINEM used for rap music does not infringe the M&M mark for candy because it is not likely to cause confusion or deception. A trademark does not give only Mars the right to use the symbol M&M. It gives Mars only the right to exclude others from making commercial use of M&M or a similar symbol that is likely to cause confusion or deception. Consumers will not confuse M&M the candy and Eminem the rapper. If EMINEM were used on candy, maybe.

What is a trademark?

Cadel starts a business making custom computers. He puts "Quill Tech" prominently on each computer sold, as well as on the packaging, his website, and correspondence. Cadel now has the trademark QUILL TECH. That does not give Cadel the sort of rights that copyright and patent give. Others can still use the phrase QUILL TECH—to discuss his business, to rhyme about pinging porcupines, to tease Cadel, to name a character in a book. . . . Trademark gives a more specific right. Cadel may prevent others from using a symbol confusingly similar to QUILL TECH in selling goods or services. It would infringe Cadel's mark to put QUILL TECH on other computers for sale at a local store. QUILL TECH or QUILT TECH on computer mice also might well infringe. The marks are slightly different but quite similar to a distinctive mark (why else use QUILL in connection with computers?) when used on similar goods. QUILT TECH on knitting needles? Not confusingly similar, so no infringement.

Why give legal protection to trademarks? It reduces consumer search costs, because buyers may rely on reputation and their past experience with the mark. Buying a computer, the consumer need not test all the components, run diagnostics on the software, test whether it meets all the specifications, and run a background check on the salesperson. Rather, the consumer may assess the reputation of the mark, such as APPLE, SAMSUNG, HP, RASPBERRY PI. That fact tends to give sellers an incentive to develop a good reputation.[2]

Trademarks are all around. Some well-known ones:

Some Famous Marks[3]

The Dogg Blogg

Declining Brands

Well he can't be a man 'cause he doesn't smoke
The same cigarettes as me.

The Rolling Stones, "(I Can't Get No) Satisfaction" (1965)

In our commercial culture, trademarks play many roles beyond their legal function. People may identify themselves by trademarks dear to their heart: a Manchester United® supporter, a Guinness® drinker, an MIT® grad, a Deadhead devoted to the Grateful Dead®, a user of Linux®, Mac®, or Windows®, a redditor of Reddit®, a Twitter® user, a wearer of Gucci® or Chanel® or Levi's®.

On the other hand, some disfavor certain brands, whether Starbucks® (to indie coffee drinkers) or Walmart® (to the snooty) or the New York Yankees® (to aficionados of the Boston Red Sox®). Marks may be evocative: Artisan's Asylum (hackerspace), Parts and Crafts (kids' hackerspace), Heart's Bend (summer camp), Wild Carrot Farm (organic farm), Boston Python Users Group (programmers, not snake charmers). Trademarks may reflect differences within a city. In London, the Ecuadoran embassy is a neighbor to Gucci, Chanel, and Yves St. Laurent; the London Hackspace abuts City Supermarket, Beds N Sofas, and Cambridge Heath Tyres.

Legally, a trademark is a symbol that indicates the source of goods or services, distinguishing them from the goods and services of others. Apple's logo on a computer indicates that the computer comes from Apple and not from other manufacturers. That indicates the source in a narrow way, by distinguishing it from other sources of computers. The Apple logo does not identify the source, in the sense that it does not tell the consumer that Apple is headquartered in Cupertino, California, or that the computer was made, under contract with Apple, in Shanghai, China. Rather, the word APPLE or the apple-shaped logo simply tell the consumer, "This comes from APPLE, not DELL, ASUS, HEWLETT PACKARD, or QUILL TECH." Likewise, Google's service mark tells the user he is not searching with BING, YAHOO!, BAIDU, or LEIT.IS. The GOOGLE mark does not say who or where Google is (a few Google searches would yield that information for most marks), or just where the servers that process Google's searches are (in fact, that is a trade secret).

Question

Suppose a factory makes acetaminophen tablets. It sells some to Johnson & Johnson, which sells them under the Tylenol® mark, and some to Walgreens, which sells them under the Walgreens®

mark. The ultimate source of the pills is the same. Can the two trademarks validly act as indicators of different sources?

Answer

Yes. A trademark acts as an indicator of source only in a limited way. It does not say who, what, why, or even where the product or service originated. Rather, a trademark simply acts to distinguish goods or services offered by one seller from those offered by other sellers. The consumer knows that the pills in the Tylenol box come from one seller and those in the Walgreens box come from a different seller. The consumer may rely on what she knows (reputation, experience, intuition) about the two sources to decide which one to buy.

Categories of marks

There are several different types of marks. A *trademark* indicates a source of *goods* (but note that "trademark" also refers to the other categories: service marks, collective marks, and certification marks). Some examples of trademarks: BOEING for airplanes, THE NEW YORK TIMES for newspapers, CROCS for sandals, SCHOLASTIC for books, ALLEGRA for an allergy medicine.

A *service mark* is a trademark for services: GOOGLE for googling, HARVEY MUDD COLLEGE for educational services, JOINT VENTURES for physical therapy services, DEATH WISH PIANO MOVERS (obvious).

A *collective mark* is used to indicate membership in a group: AMERICAN FEDERATION OF TEACHERS, FTD for an association of florists.

Question

The Department of Justice used forfeiture proceedings under racketeering law to seize the collective mark of the Mongols OutLaw Biker Gang®. Does the FBI want to sell biker jackets?

Mongol Warriors Depicted in Rashid al-Din, *Jami' al-Tawarikh* (1307)

Photograph. Britannica Online for Kids. Web. 2 Sept. 2012.

"We have filed papers seeking a court order that will prevent gang members from using or displaying the name 'Mongols.' If the court grants our request for this order, then if any law enforcement officer sees a Mongol wearing his patch, he will be authorized to stop that gang member and literally take the jacket right off his back."[4] Trademark law, however, probably would not extend that far.

A *certification mark* is different than the others. It is not used to mark the owner's goods. Rather, a certification mark signals that the goods or services meet the standards set by the owner of the mark. The (u) mark indicates that food meets specified kosher standards. NATIVE MADE signals that products meet standards set for production by Native American artists. UNDERWRITERS LABORATORY assures compliance with certain safety standards. Some mark owners test products; others use different means to ensure compliance. Often the mark owner charges a licensing fee for use of the mark in return for the certification.

Certification Marks

Are these trademarks for goods, service marks, collective marks, or certified marks?

1.

2.

(MGH stands for Massachusetts General Hospital. The full word Massachusetts is used sparingly (and then often spelled wrong) in the Commonwealth of Massachusetts. Rather, it is "Mass. Ave.," "Mass. General," "Mass. Pike," and so on.)

3.

4.

Answers

1. Trademark for electronic goods, also used as service mark
2. A service mark for medical services
3. A collective mark for membership in the Union of Auto Workers
4. A certification mark for classifying films

Marks often cross categories. Apple also sells services. MGH may sell remedies and T-shirts.

A trademark must be a source-identifying symbol

A trademark can be any symbol that signifies a source of goods or services. Qualitex's green-gold dry-cleaning pads are well known to dry cleaners. Qualitex successfully sued for trademark infringement when competitors started copying the color.[5] A symbol is not just a word, a phrase, a design, or a swoosh. The human mind is flexible. Almost anything can carry meaning. Color can be a symbol.

Qualitex's Green-Gold Dry-Cleaning Pads

Words (existing or made up) as trademarks

PEPSI, WEBLINUX, SAMSUNG, GOOGLE, ANDROID, CHANEL, OXFAM, WELLBRIDGE . . . are all examples of word marks.

Phrases as trademarks

Often words are put together to form a mark. For example, THE LOCAL BOOKIE for a bookstore, THE VILLAGE IDIOM for a local paper.

Phrases, such as slogans, can be powerful source identifiers: Nike's JUST DO IT, Burger King's HAVE IT YOUR WAY, United Negro College Fund's A MIND IS A TERRIBLE THING TO WASTE, De Beers's A DIA-MOND IS FOREVER, the U.S. Army's BE ALL THAT YOU CAN BE, and MTV's I WANT MY MTV have served the legal function of trademarks and the marketing function of motivating consumers, donors, and soldiers. And consider the fictional slogans seen on *The Simpsons*: Springfield Dog Track—"THINK OF THEM AS LITTLE HORSES"; Springfield General Hospital—"QUALITY CARE OR YOUR AUTOPSY IS FREE."

Designs (with or without words) as trademarks[6]

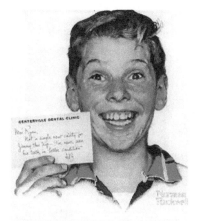

Designs (Often with Words)

The BR Logo Incorporates 31, for 31 Flavors

The Big Ten Had 11 Members at That Time, Hence the 11 around the T

An Arrow between E and X

amazon.com

The Smiley Suggests Everything from A to Z, in Zippy Fashion

Question

Any trademarks on the cereal box?

Answer

There are single-word marks (KELLOGG'S and FROSTED FLAKES); a slogan that is a mark (THEY'RE GRRREAT!); and a design mark (the image of Tony the Tiger). SIFE (Students in Free Enterprise) may also be a service mark.

How long does a trademark last?

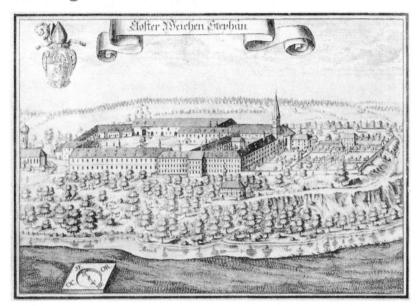

Kloster Weihenstephan (Brand since 1040)

Trademark rights last as long as the mark is used.[7] Trademarks frequently outlive patents on goods, with patents having roughly a 17-year term. Trademarks can even outlast copyright's 95 years (or life of author plus 70 years). The longevity of WEIHENSTEPHANER (since 1040), STELLA ARTOIS (1366), and LOWENBRAU (1383) attest to the loyalty of ale and beer drinkers. At the risk of anachronism, because they brought no medieval trademark infringement suits, consider the universities at Bologna (1088), Paris (1150), Oxford (1167), and Leuven (1425), all of which have long-standing service marks. Although most universities did not regularly register marks until recently, some have taken to trademark enforcement with gusto.

Trademark rights depend on continued use. The mark owner may simply stop using the symbol as a mark. The BROOKLYN DODGERS baseball team moved west, became the LOS ANGELES DODGERS, and abandoned their mark, meaning anyone could now sell BROOKLYN DODGERS gear without infringing. Many unsuccessful marks are simply abandoned, although mark owners may continue minor use to try to save a well-known brand.

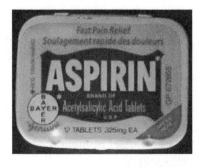

Aspirin Is Still a Mark in Other Countries

At the other extreme, a trademark may be too successful. Aspirin,[8] Cellophane, Dry Ice, Escalator, Laundromat, Mimeograph, Pilates, Thermos, Trampoline, and Zipper are all ex-marks in the United States. A trademark must be perceived by consumers as an indication of source. If one seller's trademark becomes a generic term for that category of product, then it no longer signals a particular seller.

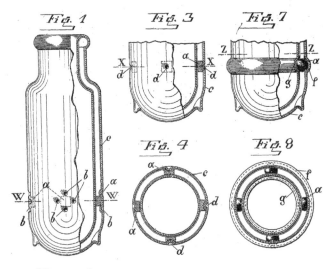

No. 872,795.   PATENTED DEC. 3, 1907.
R. BURGER.
DOUBLE WALLED VESSEL WITH A SPACE FOR A VACUUM
BETWEEN THE WALLS.
APPLICATION FILED OCT. 23, 1906.

2 SHEETS—SHEET 1.

Fig. 1   Fig. 3   Fig. 7

Fig. 4   Fig. 8

Better Known as "Thermos"

There is a well-worn track from patented invention to generic term. If the patent prevents competitors from selling similar devices, the patented product's name may become the generic term by default. A German company, Thermos GmbH, sold vacuum flasks more successfully than its competitors. Thermos GmbH obtained a patent, even though the vacuum flask was known as the Dewar bottle after its chemist inventor (as in the clerihew, "Sir James Dewar / Is a better man than you are / None of you asses / can liquefy gases"[9]). Consumers took to calling the device a "thermos," whether made by Thermos GmbH or by Widgets, Inc. The now generic term thermos was no longer a trademark. So anyone could call his product a thermos.

Questions

1. Match the patent title to the following trademarks (all of which have become generic terms in the United States):
    A. Autographic printer   Aspirin
    B. Endless conveyor or elevator   Zipper
    C. Whirligig   Dry ice
    D. Carbon dioxide in its solid form   Yo-yo
        (not patented—a product of nature)   Mimeograph
    E. Acetylsalicylic acid   Escalator
    F. Clasp locker or unlocker for shoes

2. Curious about Reykjavik, you "google" it. Is GOOGLE generic?
3. Via Mental Floss®, here are some registered trademarks. If they are not generic, what would be the generic term for each?
   a. Jacuzzi
   b. Frisbee
   c. Hula Hoop
   d. Kleenex
   e. Ping-Pong
   f. Scotch Tape
   g. Jell-O
   h. Band-Aid
   i. Velcro
   j. Novocaine
   k. TASER
4. "App store": is it a generic term or a trademark for Apple's App Store?

Answers

1. A.

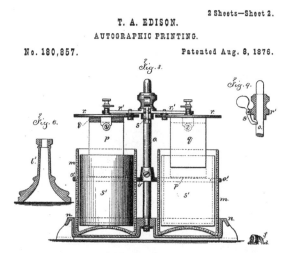

Mimeograph

B.

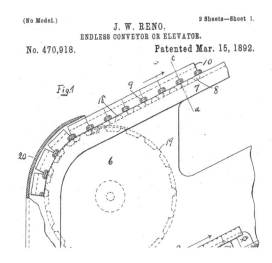

Escalator

C.

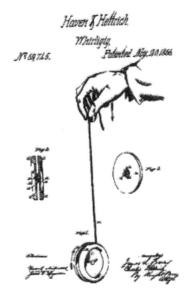

Yo-yo

___

D. Dry ice

E.

United States Patent Office.

FELIX HOFFMANN, OF ELBERFELD, GERMANY, ASSIGNOR TO THE FARBEN-
FABRIKEN OF ELBERFELD COMPANY, OF NEW YORK.

ACETYL SALICYLIC ACID.

SPECIFICATION forming part of Letters Patent No. 644,077, dated February 27, 1900.

Application filed August 1, 1898. Serial No. 687,385. (Specimens.)

To all whom it may concern:
Be it known that I, FELIX HOFFMANN, doctor of philosophy, chemist, (assignor to the FARBENFABRIKEN OF ELBERFELD COMPANY,) cause Kraut does not give the melting-point of his compound. It follows from these details that the two compounds are absolutely different. 55

U.S. Patent on Aspirin

___

F.   (No Model.)

W. L. JUDSON.
CLASP LOCKER OR UNLOCKER FOR SHOES.

No. 504,038.   Patented Aug. 29, 1893.

Zipper

___

2. Hmmm. To "google" someone or something means to use Google® to find it. That's my humble opinion—consumer surveys are more reliable. People do not speak of googling something on Bing or Baidu. But does anyone say, "I binged it" (or *bunged, bonged, or banged*, like *sing/sang*)?

   Google guards against google becoming generic. Upon Google's objection, the Swedish Language Council was kind enough not to include *ogooglebar* (*ungoogleable*) on its list of new words in 2012.[10] Or perhaps Google thinks nothing is ogooglebar.

3. Generic terms (raised eyebrow):
   a. Hot tub
   b. Flying plastic disc
   c. Ummm?
   d. Tissue
   e. Table tennis
   f. Plastic tape—or perhaps Sellotape
   g. Gelatin dessert
   h. Gauze and adhesive tape
   i. Hook-and-loop fastener
   j. Procaine hydrochloride
   k. "To be fair, 'Don't hit me with that electroshock weapon, bro!' is probably hard to shout under duress."[11]

4. App store is a new term, but it quickly became generic. The alternatives, such as "packet repository," do not seem to catch on.

What Else Would You Call It?

There are app stores, with that name, for Linux, Ubuntu, WebLinux, . . . Microsoft and others can use the term to refer to their own app store without infringing an Apple mark.

Linux App Store

Every journey begins with the first step.

This web page is the first step on the way to an App Store for Linux and BSD. An App (short for Application) in this context means a mostly proprietary Application, designed for entertainment or a special task. An App is not integrated with the underlying system.

Examples are games, ifotainment applications or an application that lets you create photo books for an printing service.

You install the Apps with the "Personal Applikation Installer" named PAPPI. This software installs an application with a single click right from the Linux App Store.

Not So Flashy

Here are some registered marks that some see as having become generic: ADRENALIN, ASTROTURF, AQUA LUNG, CROCK-POT, HACKY SACK, JEEP, JET SKI, MUZAK, NETBOOK. Others (especially the mark owners) see them as still associated with a single source.

Trademark owners sometimes take excessive measures to protect their mark for fear that they will lose their rights if they sit on them.[12] They may send cease-and-desist letters or even file lawsuits against noninfringing uses of the mark. On the other extreme, Jack Daniel's sent the "Most Polite Cease-and-Desist Letter Ever," when a book cover adapted its famous whiskey label:

> In order to resolve this matter, because you are both a Louisville "neighbor" and a fan of the brand, we simply request that you change the cover design when the book is re-printed. If you would be willing to change the design sooner than that (including on the digital version), we would be willing to contribute a reasonable amount towards the costs of doing so.[13]

Nice lawyering. The author and publisher complied.

Requirements To Be a Mark

Suppose Cadel had chosen CUSTOM BUILT COMPUTERS as the name for his custom-built computer business, putting it prominently on his products and packaging. Not long after, he notices others in the business putting CUSTOM BUILT COMPUTERS on their wares. They are not infringing his trademark. He has no trademark. A computer seller cannot trademark merely descriptive terms, such as FAST, MICRO, CUSTOM BUILT, or LOW POWER. Nor could a computer seller trademark the generic term for the product, COMPUTER, or perhaps ELECTRONIC COMPUTING MACHINE. CUSTOM BUILT COMPUTERS is a merely descriptive term—or it could be considered the generic term for the submarket of custom-built computers. To be a source identifier, a symbol must be distinctive, not generic or merely descriptive. QUILL, for example, for Cadel's custom computer business would be distinctive, and there are infinite other alternatives.

Merely descriptive terms cannot be trademarks, but descriptive terms may become so well known that they are no longer merely descriptive and instead qualify as marks: TENDER VITTLES for cat food, KOOLPAK for bags, FROSTY TREATS for ice cream.

Following is the spectrum of distinctiveness (to decide if a term is protected, ask which set it falls into):

- *Protected as soon as they are used*: Symbols that will be seen as indicators of source. An example would be QUILL for computers.
- *Not protected unless (a big "unless") they become well enough known as marks*: Symbols that will not be seen as trademarks but merely descriptive as terms; colors used as marks; product design as trade dress; surnames; and geographic marks. An example would be RELIABLE for computers (merely descriptive).
- *Never protected as a mark*: Generic terms; symbols confusingly similar to existing marks; symbols that are deceptive, functional, immoral, scandalous, or disparaging to people. An example would be COMPUTERS, which is a generic word.

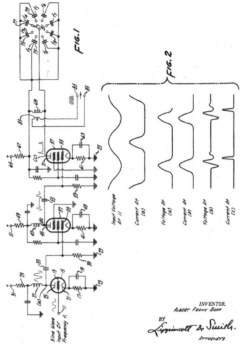

Oct. 23, 1956   A. F. BOFF   2,768,299

HARMONIC SPECTRUM GENERATOR

Filed Oct. 28, 1954

There Is a Spectrum of Distinctiveness, from Inherently Distinctive to Suggestive to Merely Descriptive to Generic

Symbols indicating source are protected

A Very Distinctive Mark for Digital Products

The word APPLE, used for computers, serves as a source identifier from the start. It does not merely describe the computer or its uses, so people will interpret it as a trademark, an indication of source. An apple seller's trademark likewise cannot merely describe apples. She could use the mark MANIAC on her apple boxes and have a trademark. The word "maniac" has no connection to apples. She could simply invent a word, as KODAK was invented for cameras. A seller can also choose a trademark that indirectly suggests the product. JOHNNY or POMPOM would be suggestive for apples.

Some distinctive marks: BILLY GOAT for a tavern in Chicago, because billy goats have no connection to the tavern business. Ditto for the BOAR'S-HEAD in Eastcheap. British pub names are often distinctive, from the tradition of choosing a name unlinked to tavernery: DOG AND DUCK, THE GOOSE AND FIRKIN, THE FLYING SPOON, THE SPOTTED DOG.[1]

Suggestive marks are a boon to language, because of the incentive for wordplay:

> JOINT VENTURES for physical therapy services
> LOCAL BOOKIE for a bookstore
> LAUNDROMUTT for self-service dog washing
> STRAIGHT DOPE for an information service

Trademark law's distinctiveness requirement often pits businesses against their lawyers. Marketers want to describe the product. But a merely descriptive term is not protectable as a trademark, so competitors can also use that descriptive term to market their products. An unrelated or made-up symbol will be interpreted by consumers as a trademark. But such a symbol tells potential buyers nothing about the product. A suggestive mark is often the sweet spot: protected as a trademark and yet offers information about the product, hopefully in a manner alluring to buyers.

Some examples from the pharmaceutical industry:

> CLARITIN and FLONASE, suggesting clarity and flow for allergy sufferers
> CARDURA, lending strength against high blood pressure
> REQUIP, fighting against Parkinson's disease
> AMBIEN, STILNOX, and LUNESTA, as respites from insomnia
> PROVIGIL, for sleepiness
> ZOLOFT, WELLBUTRIN, and PROZAC, to lift from depression
> ABILIFY, for treating schizophrenia
> IMURAN, used as an immunosuppressant

With drugs, suggestiveness comes into play not just as a marketing tool. The user's mental state can affect how well the user considers the product to work. A suggestive name could have a stronger placebo effect than an outright descriptive name.

Generic terms are not protected

To use a classic example, an apple seller cannot have a trademark in the generic term for the product, apple. Suppose one apple seller could prevent others from using the word "apple." That would defeat the purpose of trademark law, which is to make it easier for sellers to communicate with buyers and thereby reduce search costs of buyers. Other sellers would have to find other ways to describe the apples. "For sale: Round fruit of a deciduous Eurasian tree." An apple seller can certainly use "apple" to sell the goods, but he cannot get a trademark on the word.

"APPLES" Is Generic for Apples, although Distinctive for Digital Products

Arthur Rothstein, Farm Security Administration photographer

Question

Trademark infringement?

v.

Answer

No. "Duck tours" is the generic term for city tours in World War II surplus amphibious vehicles, so there is no trademark to infringe.

Merely descriptive terms are not protected

There is no trademark protection for merely descriptive terms. If one apple seller had a trademark in JUICY, RED, or SEEDLESS, that would obstruct other sellers. Moreover, like generic terms, <u>merely descriptive terms do not serve as source identifiers</u>. If a consumer sees "Juicy" or "Red" on a box of apples, he will likely think those words merely describe the product and that those words might be used by anyone trying to sell apples.

1. Of the following marks from *The Simpsons*, which would be merely descriptive and therefore not trademarks?[2]
    > Pay & Park & Pay (parking lot)
    > Quetzal's Pretzels (pretzels)
    > What They Eat in Iceland (restaurant)
    > Compu-Global-Hyper-Mega-Net (Internet service provider)
    > *Fretful Mother* magazine (for helicopter parents)
    > The Grateful Gelding (horse stable)
    > The Legless Frog (restaurant)
    > Tuition Academy (private school)
    > High-Pressure Tire Sales
    > Just Take Me Home Taxi Cab Co.

2. Is this mark merely descriptive for a note-taking app?

Answers

1. The answers are mixed:
    > Pay & Park & Pay—quite descriptive
    > Quetzal's Pretzels—"pretzel" is generic, but "Quetzal" is distinctive for pretzels, so the mark is protected, but other pretzel sellers can still freely use the word "pretzel"
    > What They Eat in Iceland—descriptive
    > Compu-Global-Hyper-Mega-Net—the individual terms are merely descriptive, but stringing them together is distinctive
    > *Fretful Mother* magazine—descriptive
    > The Grateful Gelding—suggestive, thus protected as a mark
    > The Legless Frog—likewise suggestive
    > Tuition Academy—merely descriptive, if doubly so
    > High-Pressure Tire Sales—ditto
    > Just Take Me Home Taxi Cab Co.—descriptive

2. Elephants never forget, they say. The design would be suggestive, but not descriptive for note-taking apps. Other note services do not need to use pictures of elephants to describe their services.

Well-known descriptive terms may become trademarks

You may have noticed this chapter keeps saying "merely descriptive" terms are not protected. Why not just say "descriptive" terms are not protected? Because descriptive terms can become source indicators.

"Coca-Cola" was a merely descriptive term (by way of example—the law was different in 1904). It described the key ingredients: coca leaves and kola nuts. The drink became popular. Before long, the primary meaning of "Coca-Cola" became an indication of source that differentiated it from other sources of beverages, such as CANADA DRY. Now COCA-COLA is one of the most recognized trademarks in the world. Once a descriptive term becomes distinctive, it is protected as a trademark. It serves as a source identifier, and consumers would be confused only if other sellers were allowed to label their beverage with the same name. In the case of COCA-COLA, other sellers may still use the descriptive term "cola" to describe their product. Moreover, RC Cola may list coca leaves and kola nuts as ingredients, even prominently, without fear of infringing the COCA-COLA mark.

Another merely descriptive term that became distinctive is MICROSOFT®. When Bill Gates started a little company selling software for microcomputers, he logically named it Microsoft. The term was merely descriptive, and so he had no trademark. Not that he paid much attention. Many start-up businesses have other things on their mind than qualifying for a trademark. Once MICROSOFT became well-known, the term was no longer merely descriptive. It had become distinctive and so was a protected trademark. Although WINDOWS was a famous MICROSOFT brand, some have questioned whether it was merely descriptive—or even generic. Most businesses, however, do not become so well known that merely descriptive names become distinctive. They are better advised to pick a distinctive term from the start, like APPLE for computers. That worked out pretty well—and, incidentally, also gives some insight into the difference in philosophy between the two companies.

Normally, a product's color would not qualify as a mark. The color is not an indication of where the product came from but is used simply to make the product look nice (white shirts or a black couch) or sometimes to fit the environment (blue on boats) or function well (dark solar panels absorb light better than pale ones). Christian Louboutin, however, has a trademark on the color red as used on the soles of his shoes, in which the uppers are a different color. He used that unusual color for shoe soles so consistently that red soles literally became his trademark—like the green-gold color of the dry-cleaning pads for Qualitex.

Trademark Registration 3,361,597

Distinctive designs may become trademarks

Like a descriptive term, some other symbols may become trademarks. Normally the shape of a product, like a penguin-shaped cocktail shaker, would not be interpreted as a mark. When the picture appears in advertising or on a package, it would not be interpreted as an indication of source. But for some well-known products, product design may become a trademark. The Coca-Cola bottle's shape became so well known as to qualify as a mark.

Fig. 1

Nida's Penguin

Norpro's Pete, Cocktail Shaker (from Advertising)

Surnames may become trademarks

A surname, like Dent, will normally be seen as simply someone's name. But MCDONALD'S is so well known as a mark for fast food that it has become distinctive, hence a trademark. Geographic terms may likewise become distinctive, like KENTUCKY FRIED CHICKEN (KFC).

Questions

1. Can Boise State's blue artificial turf in its football stadium be a trademark?
2. A law school publisher's line of red-and-black books are well known to teachers and students, readily distinguished from other publishers. Red and black have been used by many others, not to mention on Stendhal's 1830 novel *The Red and the Black*. Can the combination of red and black be distinctive enough to be a trademark?

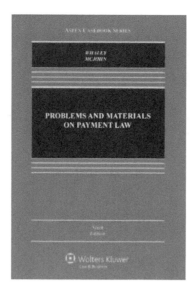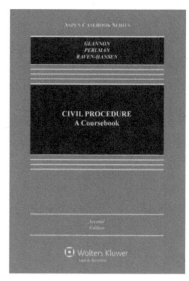

Answers

1. Yes, if it becomes well known enough to become associated with Boise State. Well-known colors, especially combinations, for sports teams and schools may also qualify as trademarks.
2. Yes, when law teachers and students recognize the publisher simply by the familiar red-and-black covers, those colors have become a trademark.

Some symbols cannot be trademarks

As has been explained, generic terms cannot be trademarks. Trademark law likewise denies protection to certain other symbols, such as those that are confusingly similar to existing symbols; those that represent functional matter; those that are disparaging to people, immoral, or scandalous; or those that would be deceptive.

Symbols confusingly similar
to existing trademarks

If BLUE FIN TECHNOLOGIES is a mark for data analysis software, then another company cannot use BLUE FINN TECH as a trademark for database software. A symbol cannot be a mark if it is confusingly similar to an existing mark. But a restaurant could have the trademark BLUE FIN. That mark is similar to the mark for the data analysis software—but not confusingly similar because of the differences in products.

A trademark lawyer queried a listserv for "examples of well-known similar marks that peacefully coexist in commerce" and quickly got back the following:

AMERICAN _____ (fill in the blank)
CANON (cameras) and CANON (textiles, towels, sheets)

CLUE (board game) and CLUE COMPUTING
EAGLE shirts, EAGLE pencils, EAGLE pretzels, EAGLE condensed milk, EAGLE hardware, EAGLE insurance, EAGLE bicycles

LOONEY TUNES is a famous mark for cartoons and also the mark for a used music store. WATSON, named after the founder of IBM, is an artificial intelligence system that beat humans at *Jeopardy!* Another WATSON, named after Alexander Graham Bell's assistant (the first telephone call: "Come here, Watson, I need you"), is a computer speech system.[3] BBN developed the ARPANET and the first Internet router. BB&N is a nearby school. IRON MAN for a triathlon coexists with IRON MAN for comics and movies.

But, remember, <u>a symbol cannot be a mark if it is confusingly similar to an existing mark</u>. Cadel could not use APPLE for computers—or any other confusingly similar symbol, such as APPEL, APPPLE, or perhaps even MACK.

Questions

Confusingly similar?

1. Could RUMBA be a mark for exercise classes, if ROOMBA® is already in use for vacuuming robots?
2. Is deCode for genetics analysis confusingly similar to dEcode for "computer software for use in association with an educational reading program"?
3. Could Cadel use SAMSUNG as a mark for custom-made computers?
4. M&MC is a conference where physicians discuss errors contributing to morbidity and mortality. Could the conference register M&MC as a service mark, or is it confusingly similar to the existing mark of M&M for candy?
5. Will the real PENGUIN® please stand up?

6. Hawk or dove?

 v.

7. The sky's the limit?

 v.

8. Spamalot?

 v.

Answers

1. Similar marks may coexist if not confusingly similar. ROOMBA for vacuuming robots and RUMBA for exercise classes are not confusingly similar. One mark is for a labor-saving device, and one for a labor-encouraging program, to a dance beat. No one looking for ROOMBA will accidentally choose RUMBA, or vice versa. (But perhaps RUMBA for dance classes might not be distinctive, because "rumba" is a term for a genre of Cuban ballroom dance.)
2. Almost identical, but quite different markets, so not confusingly similar.
3. Cadel sells custom-made computers, and Samsung sells computers that people use as telephones, to text, and to Twitter. Different, but confusingly similar. Fortunately, Cadel has infinitely many other choices.

4. Not confusingly similar.
5. They can all be marks for different markets. This would be concurrent use, as lawyers call it.
6. Similar sized bars of white stuff, but they are for different markets. They don't sit on the same supermarket shelf, and no consumer will think the soap maker has branched into ice-cream sandwiches. Not confusingly similar.
7. The same word is used for beverages, but it is presented differently, with a different design and label, and the two products are for quite different markets. Not confusingly similar.
8. SPAM the processed meat begat "Spam" the Monty Python sketch ("spam spam spam spam spam spam spam"), which begat "spam," the word for unwanted mass e-mails, which begat the tech company SPAM ARREST. But buyers will not think that Hormel, the maker of SPAM the meat, has developed e-mail security software.

    However, Hormel has shown good humor in licensing an image featuring *Spamalot*, the musical.

Symbols for functional matter

A road-sign maker may be known for the unusual design of its flexible temporary road sign, handy for construction sites. But others may copy the sign's design. Functional matter cannot be trademarked; otherwise, one could in effect get perpetual patent protection by using trademark law.

The Design of a Road Sign Was Functional and So Not a Trademark, Meaning Competitors Could Copy the Design

Functional design is involved for a toilet plunger, so there is no trademark to infringe.

Functional Design Not Trademarked

1. Would the vertical opening motion of a Lamborghini car door be a trademark?

2. "The red ones calm your nerves. The white ones relieve your headache. The blue ones are for asthma."

   "Amazing. Such little pills, and each one knows exactly what to do."[4]

   When the patent expires on a drug, generic manufacturers will sell substitutes. In order to win over purchasers, a generic seller may wish to sell the drug in the form familiar to patients: the same shape tablet, in the same color. Can the generic seller parallel the previously patented pill's preferred pattern and pigment?

3. Can Secalt prevent others from copying its traction hoist if its distinctive cubist look is well known?

Answers

1. Under U.S. law, the way a car door opens is functional and therefore cannot be protected as a trademark, even though Lamborghini's movement is distinctive.[5] To prevent copiers, Lamborghini would need a patent. But other countries may allow such marks.

2. A drug's chemical function may not depend on the shape of the tablet or its color. But because of the strong mental association the consumer has with the familiar tablet, and the consumer's suspicion that a different-looking tablet will not work as well, the shape and color could have become functional because they affect how effective the drug will actually be for the consumer.[6]

3. The look may be distinctive, but it is functional and so cannot be protected as a trademark. However, if they painted the bottom red . . . .

Symbols disparaging to people or immoral or scandalous

Scandalous symbols such as obscenities cannot be trademarks. Not every objection will make a mark scandalous and thus nonprotectable. In simpler times, in households of General Motors' workers, the word "Ford" was often considered a swearword (albeit a mild one). Sometimes children would breathlessly report to their parents that they had seen a car with that swearword emblazoned on its hood.

"Our Prices Are Sofa King Low": Scandalous

Rejected as Scandalous

Question

Is the mark for WASHINGTON REDSKINS invalid as being disparaging to a group of people?

Opinions vary strongly on whether WASHINGTON REDSKINS®, for a professional football team, is invalid as being disparaging to Native Americans. What do you think?

Symbols that are deceptive

Trademarks are supposed to help consumers, not trick them. <u>A trademark need not be literally true. But a symbol that deceives consumers as to something that affects their buying decisions is deceptive and cannot be a mark.</u> BREATHASSURE for ineffective breath-freshening mints, for example, would be deceptive.

Questions

1. Despite the umlauted Nordic-looking name, the ice cream comes from New Jersey. Deceptive?

2. Mom & Pop Hardware on *The Simpsons* is actually "A Subsidiary of Global Dynamics, Inc." Deceptive (if it was a real business)?
3. SWEDISH FISH is candy, not fish, and not from Sweden. Deceptive?
4. SUB ZERO: deceptive mark for refrigerators? The temperature in the refrigerator is cold but above freezing, while the freezer is at freezing, not below freezing.

Answers

1. Unless consumers think the ice cream is from Scandinavia, *and* unless they care about where the ice cream comes from, they are not being deceived. If they would buy it anyway, there is no deception.
2. Not unless consumers indeed shop there to support a local family business. (Anyway, it is not a trademark because it is a merely descriptive or even generic term.)
3. Admittedly, as literally read, this is false—but it is not deceptive. Buyers know they are getting candy, not dried herring.
4. Likewise, this is false—but not deceptive. Consumers know what a refrigerator does.

Choosing, Registering, and Owning a Trademark

Ownership

Steve Jobs started Apple Computer, but he did not coin the word "Apple." Nor was he the first to use APPLE as a trademark. Businesses had been named Apple before, such as Apple Records, started by the Beatles. So how did Jobs's company come to own the trademark APPLE for computers?

A trademark belongs to the first person in a market to use it. A business can establish trademark rights simply by using the mark on goods when they are sold or transported. There is no special rule as to placement of the mark. It must simply appear to consumers as a mark. Apple first sold its laptops with the picture of the apple oriented toward the user when closed. This meant that the logo would be upside down, from the point of view of others, when the laptop was in use. Eventually, it was flipped. That had no effect on trademark rights. The trademark can be used on the packaging. For some goods, like donuts or nanotechnology, putting the mark right on the goods is not so practical.

The trademark owner need not originate the mark. The owner need only control the first use of the mark. Existing words and phrases may be adopted as a mark (FACEBOOK, TWITTER, LANDS' END, YOUNG MONEY, LEGAL SEAFOOD). The mark may be invented by the owner, an employee, or anyone. Linus Torvalds founded Linux®, the Unix-like open source operating system. Linus chose the mark FREAX (from Free, Freak, and X). But the system administrator of the first server to host the code disliked the name Freax and used the folder name Linux.[1] Some trademarks were thought up as nicknames by the public and adopted by the company: BIG BLUE for IBM, MARKS AND SPARKS for Marks & Spencer, KFC for Kentucky Fried Chicken, and COKE for Coca-Cola.

Question

> An employee of the Baltimore Ravens professional football team, in his spare time, designs an avian logo and suggests the team use it. They do but fail to give him credit. Trademark infringement?

Answer

> No trademark infringement, because the employee has no trademark. As one lawyer wryly put it, "He doesn't own a football team."[2] But he may own the *copyright*. . . .

Two "Dawn Donuts"

The first user has rights to the mark in the area of actual use. Where one party used DAWN DONUT as a mark in Michigan, and the other used it in New York, both had valid marks in their respective areas of actual use. THE BURGER KING operated in Mattoon, Illinois, before titan BURGER KING registered its mark. The Mattoon Burger King had the mark in Mattoon.

To establish nationwide rights, one must use the mark nationwide—or register the mark. Registration is not required to have trademark rights, but it makes the mark stronger quicker.

Questions

1. Amazon.com began selling books online and registered the AMAZON.COM mark. Unbeknownst to Amazon founder Jeff Bezos, a women's bookstore in Minneapolis had been operating for years under the unregistered mark AMAZON BOOKSTORE. The names are slightly different, as are the markets (in person versus online book sales). But all things considered, the marks are confusingly similar. Who gets to keep the mark?

Little Store Beats Big Store, but Only on Its Home Turf

Jenny Chandler (1890–1910), http://thehenryford.wordpress.com/2009/06/19/
http://commons.wikimedia.org/wiki/File:Kuznetsky_Most,_Daziari_bookstore_1900s.jpg

2. EMINEM® is a famous mark for rap music, since around 1999. A new rapper calls herself M&M and starts selling concert tickets and recordings—with the blessing of Mars, Inc., holder of the M&M mark for candy since 1941. Who is infringing on whom (or whom&m)?

1. Amazon Bookstore wins in Minneapolis. Amazon.com gets the rest of the country by means of registration. They would split the country, albeit unevenly. You won't be surprised to learn that the parties settled the lawsuit, with the owner of Amazon Bookstore shortly thereafter contentedly retiring.
2. Mars's M&M mark is a mark for candy wrappers, EMINEM is a mark for rappers. M&M, the runner-up rapper, would infringe EMINEM®.

Choosing a name[3]

1975 Version

Many famous marks were chosen without too much thought. FORD MOTORS simply used the name of the founder, Henry Ford, as was the case with Kiichiro Toyoda, Karl Benz, and the brothers Renault (but not General Motors). The brothers McDonald sold hamburgers. M&M came from the last names of Messrs. Mars and Murrie. Other marks came straight from the product. MICROSOFT could have been GATES & ALLEN (or AGATE), but "Microsoft" merely described what they sold: microcomputer software, like BASIC. APPLE was more creative, but it was chosen by Steve Jobs rather than market research. Jobs reportedly picked apples in the summer and liked the Beatles' music, which he associated with Apple Records. FACEBOOK came from facebooks issued in college dorms. VELCRO? A portmanteau, from the French words *velours* and *crochet*, velvet and hook.

Other businesses take great care in choosing a mark. BLACKBERRY, for e-mail devices, was chosen after using mind maps of hundreds of words, many hours of brainstorming and winnowing candidates, and consultation with linguists. RIM, the maker of the device, had originally favored MEGAMAIL, but that was thought to trigger fears in users of being inundated with e-mail. BLACKBERRY was so successful that others tried (in vain) to register the mark CRACKBERRY.[4] But whether the mark, as opposed to the product, helped sales cannot be determined. Certainly, it did not save BlackBerry from rapidly disappearing in the wake of the iPhone and Android.

Ford Motor Company, when choosing a name for a highly touted car in the 1950s, had an advertising agency come up with names like Corvair and Ranger, but rejected them after market testing. Ford turned to a notable poet, who suggested "a list of names that demonstrated a serene distance from the commercial marketplace: among them were Intelligent Bullet, Utopian Turtletop, and Mongoose Civique" (to quote *The New Yorker*[5]). Finally, Henry

Ford simply chose the first name of his son, Edsel Ford—and EDSEL has become synonymous with a flop product.

In choosing a name, a business will wish to avoid mystifying, confusing, or deterring its potential customers, investors, and partners. The quickly famous mark of iPad was able to overcome two speed bumps. Half the U.S. population thought the new product sounded like a feminine hygiene product. In China, IPAD was already a registered mark—which Apple was eventually able to buy for some $60 million. Other marks with less momentum have run aground with "cross-linguistic gaffes,"[6] like CREAP coffee creamer from Japan, BUM potato chips from Spain, and in Mexico, the CHEVY NOVA— in Spanish, the "Chevy no go." Ensuring that a suitable domain name is available for the product's website is also key, which is one reason many businesses now invent words as their brand.[7]

WHOLLY AROMATIC CARBOCYCLIC POLYCARBONAMIDE FIBER HAVING ORIENTATION ANGLE OF LESS THAN ABOUT 45°

Better Known as Kevlar®

Vetting a mark

The most common pitfalls are that the proposed symbol might be confusingly similar to an existing mark or that the symbol is merely descriptive. To avoid the first problem, a trademark search should be made for existing marks. To avoid the second, one should evaluate whether the mark is merely descriptive (or functional, scandalous, or deceptive, as discussed in Chapter 15) and if so, change it.

Trademark search

> I did a trademark search on all of the excellent product
> names you suggested.
> Every one of them is taken.
> As are the not-so-great ones. And iCrud, eatdirtanddie,
> and defectiveproduct.
> So our new product name will be a combination of grunts
> and shrieks.
>
> Scott Adams, *Dilbert: This Is the Part Where You Pretend to Add Value*

A trademark search can check if similar marks are already in use. Trademark searching is more of an art than a science. Suppose Ford is considering the name Mongoose Civique for a new model of car. Ford would certainly search the trademark register for "Mongoose Civique." If it is already in use (perhaps the spurned poet used it for her own products), Ford could decide not to use the mark. MONGOOSE CIVIQUE for cars would not likely be confused with MONGOOSE CIVIQUE for poetry recordings. However, even though the relevant products are wildly different, the mark is so distinctive that consumers aware of both might think they are related. If Ford really liked the mark, Ford could negotiate to buy the rights from the poet.

A Mongoose Mark

Even assuming that MONGOOSE CIVIQUE is not registered, Ford's search has just begun. MONGOOSE by itself is registered for such diverse offerings as animal-activated pet feeders, volleyball game–playing equipment, and cylinder heads for engines, not to mention MONGOOSE METRICS, JANE'S MIGHTY MANGO MONGOOSE CHOCOLATE CHUNK COCONUT MACADAMIA NUTS, and MONGOOSE ULTRA STORM (for bicycles). Ford would have to go through all the 100-plus items to make sure none were for cars or confusingly similar products or services. Then Ford would have to decide whether all were distinct enough from MONGOOSE CIVIQUE for cars. Only the last issue presents a possible problem. Would MONGOOSE CIVIQUE for cars be confusingly similar to MONGOOSE ULTRA STORM for bicycles? Both involve transportation, but they are different products and are in quite different markets. The answer is probably no.

Literally searching the register, however, is not enough. MONGOOSE CIVIQUE can be preempted not just by identical marks but by marks similar in spelling, meaning, sound, or some other way. Ford should search for words with similar meaning. "Viverrine" is the family of animals that includes the mongoose, but it has not been registered as a mark (opportunity knocks!). The civet is in the same family—and CIVET has been registered for financial analysis services and for coffee, neither of which would be confused with MONGOOSE for cars.

Next, words with a similar sound, such as MANGOES, huMONGOUS, AMONG US, and (for CIVIQUE) SKIVVY, CEVICHE, and CIVIC, should be looked at. The last one presents issues: civic is similar in both sound and meaning to CIVIQUE. MONGOOSE CIVIQUE sounds not unlike HONDA CIVIC, though not confusingly similar (we might think it is not confusingly similar—Honda might not agree). But Ford might not use a mark that is even vaguely like HONDA CIVIC for marketing reasons. Ford might prefer a more distinctive mark.

Suppose Ford, after searching with great imagination, is satisfied no confusingly similar mark is registered. The search has miles to go before it sleeps. Ford must search the world for confusingly similar unregistered marks. For such purposes, trademark search companies keep databases of trade names, corporate names, and product names. But simply googling "Mongoose Civique" is a good start:

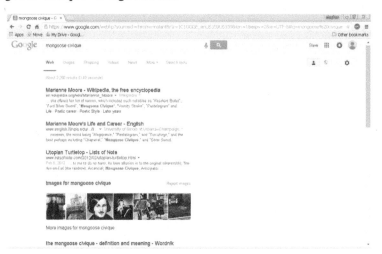

Search Results for "Mongoose Civique"

The most likely results from a Google search related to Ford's flirtation with "Mongoose Civique" are pages about the poet who thought up the name, along with a keyword ad from Walmart for the Mongoose Ultra Storm bike, mentioned earlier. A searcher could then google various words similar in sight, meaning, or sound. Assuming Ford is interested in selling the car abroad, it could pay to search foreign trademark registries as well. No trademark search can be perfect, but Ford may conclude at some point that the chances are about nil that a mark exists that is confusingly similar to MONGOOSE CIVIQUE for cars.

Mongoose Image Search Results

Recall that a mark can be any symbol. It can be a design, like Nike's swoosh, or a sound, like MGM's lion's roar or Dilbert's combination of grunts and squeaks. Searching such marks is difficult but may become easier. Google can search by image. Searching with a drawing of a mongoose that was registered as a mark for sports equipment did indeed find that drawing on other sites (along with, again, Mongoose bicycles), presumably because the image file was tagged with the word "mongoose."

But visual and sound searching lags behind searching for words and numbers, which come organized in standardized, digital form. For example, a search with the image of a lamp base statuette from a copyright case thought it resembled the first computer programmer, Ada Lovelace, former vice president Fritz Mondale, and Chairman Mao. Sensible algorithm: people usually search for people, not lamp bases.

Image Search Results for Statuette

Evaluating the mark

As mentioned earlier, in addition to a trademark search, the examiner will consider whether the mark is merely descriptive or if it is functional, scandalous, or deceptive, any of which problems will cause a rejection of the application for registration. Therefore, in addition to a trademark search, the owner of the trademark needs to evaluate it for these other problems, which are discussed in detail in Chapter 15.

Benefits of registration

Like copyright, trademark registration is not required but does give considerable benefits. Registration counts as nationwide use. It may make it easier to stop infringers and make them pay more money. Registration makes it easier for the owner to prove that the mark is valid.

Registration also permits the use of the ® symbol after the mark, such as in PIRATES®. Confusingly, there is a "supplemental register" for registration of symbols that could become marks, such as merely descriptive brand names that may become well-enough known. One can also use ® by those symbols, so ® does not necessarily mean the symbol is a valid trademark. If a mark is not registered, the mark owner can use ™ for trademarks or ˢᵐ for service marks.

The registration process

Along with a fee ($275 and up), the application requires specimens or facsimiles of the mark as used in commerce. An examining attorney at the USPTO considers the application. He may deny it for any of the reasons discussed in previous chapters. The two most common reasons are that the mark is merely descriptive or is confusingly similar to an existing mark. He might also deem the mark generic, immoral, deceptive, scandalous, or functional. If he finds the mark registrable, it is published in the *Official Gazette of the Patent and Trademark Office*. Others may then file an opposition proceeding. To guard their marks, companies review the *Gazette* regularly, or hire a service to do so, and may file oppositions to marks they do not like. That could be a mark similar to their own mark. It could also be a filing by a competitor that they think is deceptive, or merely descriptive. If an opposition is filed, the USPTO hears both sides and decides.

If the applicant makes it past the examining attorney and any opposition, the mark is registered (for another fee, $100 or more). It may be renewed every ten years. If the application is denied, the applicant may appeal, first to the USPTO and then to the courts.

Trademark Infringement: Confusingly Similar?

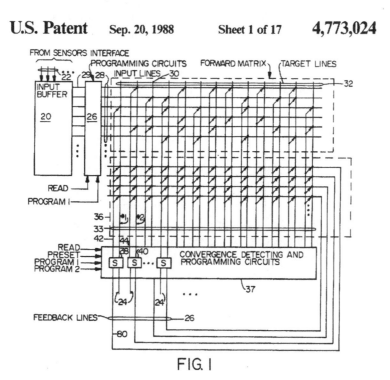

US. Patent   Sep. 20, 1988   Sheet 1 of 17   4,773,024

FIG. 1

Synaptics, Inc.'s "Brain Emulation Circuit with Reduced Confusion" (Confusion Will Be a Thing of the Past)

What is important is to spread confusion,
not eliminate it.

Salvador Dalí
(engineers and trademark lawyers may differ)

Would this be trademark infringement? A candy store sells bags labeled M&N that are filled with generic candy. No, that would not be trademark infringement. How about this? Van Halen's contracts specify that the rock band will be provided with bowls of M&Ms after the concert in which all the brown M&Ms have been removed (not a case of being prima donnas here but done only to see whether the promoter had read the contract, which had clauses that needed reading, such as the considerable electricity and structural requirements).[1] Van Halen can use the M&M® mark as much as it wants, except to sell confusingly similar goods or services.

Also not trademark infringement: Reese's Pieces. These nuggets compete with M&Ms but are not sold under a mark confusingly similar to M&M®.

Most mentions of trademarks (and similar symbols) are not remotely infringing. To contrast, copyright allows one to copy, adapt, publicly distribute, perform, and display the work (subject to fair use). Trademark gives a much narrower right: *only* to prevent commercial use of the mark or a similar symbol that is confusingly similar.

Commercial use[2]

To infringe, the use of a trademark must be commercial in the sense of connection with the sale of goods or services. Normally, mentioning a mark in poetry, song, a book, or conversation does not infringe. It was not trademark infringement when a Swedish couple named their child "Oliver Google Kai."[3] If Oliver grows up and starts up a search engine company, however, he may not be able to name the company after himself. That would be a commercial use. So, possibly, would using a mark in an advertisement in a book or embedding an advertisement in a music video be commercial use. If the use had no artistic relevance or was expressly misleading, courts have ruled it could be infringement.

Questions

1. Alrightnik carries around a knockoff Gucci purse. Is Alrightnik infringing GUCCI®?
2. Protest signs accuse BP®, EXXON®, and SHELL® of various environmental wrongs. Trademark infringement? Would it matter if the signs made baseless claims?
3. To discourage burglars, a homeowner leaves a sign in the window, "PROTECTED BY ADT® SECURITY SYSTEMS." The sign is a useful relic of a long-lapsed contract. Trademark infringement?
4. Some examples of music with marks, via *The Straight Dope*: "got your Wayfarers® on"; "Jose Cuervo®, you are a friend of mine"; "She keeps Moët and Chandon® in a pretty cabinet"; "Now look at them yo-yo's [generic], that's the way you do it, you play the guitar on the MTV®"; Frank Zappa: "Is that a Mexican poncho? Or a Sears® poncho?"; Lou Reed's "Walk on the Wild Side" refers to Valium®. Some more contemporary cuts use trademarks to evoke luxury: Dom Perignon®, Air Jordans®, Cartier®, and various Gulfstream jets, as in "I know G4® pilots on a first name basis" and "Fly like a G6®." Does it infringe to mention a mark in a song?

5. The no-commercials British Broadcasting Corporation (BBC) formerly had a policy against playing songs mentioning trademarks. That reportedly led Mott the Hoople in "All the Young Dudes" to alter "stealing clothes from Marks and Sparks®" to "stealing clothes from unmarked cars" and the Kinks in "Lola" to switch "Coca-Cola®" to "cherry cola." Did the BBC fear trademark lawsuits?

6. Did pop artist Andy Warhol infringe the mark of Campbell's Soup®?

7. In an immersive video game set in a fictional city, players can walk into an Amazon bookstore (looks just like the website) and buy books and other goods, which are shipped to them in the real world. But the game Amazon is not authorized by Amazon.com. Unbeknownst to the players, the stuff is bought from the game company. Trademark infringement?

Answers

1. Whoever sold the knockoff certainly infringed. But Alrightnik's use is not commercial, so Alrightnik is not infringing.
2. Use of a mark can infringe only if its use is commercial. Expressing opinion is not trademark infringement even if wrong. There's

no likelihood that consumers would think that the oil companies sponsored the protests. Nor would even baseless claims be false advertising in this context.

3. The homeowner is not selling goods or services, so there is no trademark infringement. It may be breach of contract.

4. None of those uses would infringe. Using a mark within a creative work is not commercial unless it is linked to the sale of goods or services (such as in app sales). And even if it was commercial, it is not confusingly similar. Buyers will not think that Sears sponsored Zappa. But use of a trademark with no artistic relevance or of one that is expressly misleading could infringe. This point applies to the other questions as well.

5. Using a mark in a song, as opposed to commercials, which the BBC did not have, does not infringe. There are other reasons that the BBC might have for its policy. Such "product placement" may help the product, even when not complimentary, on the theory that all publicity is good publicity. The BBC may simply have wished to steer clear of promotion of brands.

6. Same answer as using a mark in a song: no infringement. Some would say it is because the use is not commercial, but others would say that Warhol was selling art, so it is commercial. But there is still no infringement because typical buyers would not be confused as to whether Campbell's authorized the art.

7. Now, *that* would be a commercial use that is confusingly similar, meaning it would infringe the AMAZON.COM® mark. People are likely to think they are dealing with Amazon.com. The expressly misleading commercial use would infringe.

Confusingly similar possibilities

> Peace, ho, for shame!
> Confusion's cure lives not
> In these confusions.
>
> William Shakespeare, *Romeo & Juliet* (Friar Lawrence, on the star-crossed lovers' fate)

Even commercial use of a mark or similar symbol underlines only if it is con- fusingly similar. This is a murky standard, because mental life in a consumer economy is one great blooming, buzzing confusion.[4]

To discover infringement you must look to any relevant factor (remind you of fair use in copyright law?).

Similarity of marks

The marks need not be identical to infringe. PEOPLE'S FRONT OF JUDEA™ would infringe the JUDEAN PEOPLE'S FRONT®. They need only be confusingly similar. The more similar the symbols, the more likely buyers will confuse the two. Similarity can be in sound, sight, or meaning.

Confusingly similar?

1. Database service versus car maker.

2. Johnny Love Vodka versus Pucker Vodka: Lips are sealed?

3. WAGAMAMA (Japanese restaurant) versus RAJAMAMA (Indian restaurant).

4. Little sci-fi conference organizer Dreamwerks versus instantly famous movie studio DreamWorks:

5. United States Post Office (USPS) versus United Parcel Service (UPS):

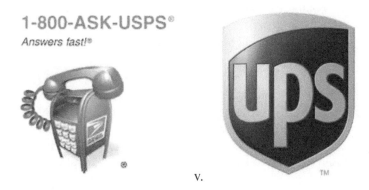

6. Two banks: Peoples versus People's (only an apostrophe to differentiate them):

 v.

7. Christian Louboutin versus Yves St. Laurent:

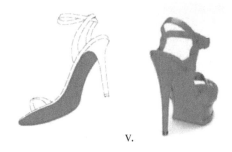

v.

8. In the early days of personal computers, APPLE GALAXIAN was an unauthorized version of the video game Galaxian, made to run on Apple computers. Infringement?

Answers

1. Twin Lexuses, but quite different markets, so not confusingly similar. Ask yourself: would someone seeing a Lexus think that Lexis the database company now sells/sponsors cars?
2. Both have lipstick prints on vodka bottles, but the products are easily distinguished.[5]
3. Opinions differ. RAJAMAMA was deemed to infringe WAGAMAMA in Benelux (not to be confused, as it were, with luxurious Benetton), but not in England.[6]
4. When DreamWorks (a movie studio formed by famous executives) started up in a related field, some people would likely assume that little Dreamwerks (the sci-fi conference organizer), which actually came first, was a knockoff outfit. DreamWorks therefore infringed by reverse confusion.
5. USPS and UPS confuse your author regularly, but they seem to peacefully coexist.
6. *People's* is such a common term in banking that these two are not confusingly similar.
7. An all-red shoe is not confusingly similar to a line of red-soled shoes with contrasting uppers.[7]
8. APPLE GALAXIAN sounds like it was from GALAXIAN or APPLE, and so had the distinction of infringing two marks at once. Nobody apparently cared. APPLE GALAXIAN made a mint.[8] "In the good old days of trade, in which our forefathers plodded on . . . ."[9]

Distinctiveness of the mark

The VIRGIN family of marks (VIRGIN ATLANTIC, VIRGIN MONEY, VIRGIN MEDIA, VIRGIN TRAINS, VIRGIN MOBILE, VIRGIN MUSIC) are quite well known and distinctive. Few other marks have the word *virgin* in the name. Using VIRGIN on consumer goods or services in any related field would likely confuse consumers. Someone who saw VIRGIN LUGGAGE might well think that VIRGIN ATLANTIC had branched out to that related market. By comparison, UNITED AIRLINES is well known but less distinctive: UNITED is used all over the place by other businesses. UNITED used on luggage would not be likely to confuse.

Microcenter, a chain of electronics stores, does not infringe the famous MICROSOFT mark. The marks are somewhat similar. Microsoft does operate retail stores, called Microsoft stores. But *micro* and *soft* are words used everywhere in the tech field. Consumers are not likely to think a store named Microcenter is somehow sponsored by Microsoft.

On the other hand, someone who opened an Apple Center to sell computers and other tech would be infringing the mark of Microsoft's rival, Apple®. APPLE is highly distinctive for technology. In this field, the word *apple* has relatively few uses other than to refer to Apple products (although *application* and its alter ego *app* and cousin *applet* are pervasive).

The more distinctive a mark is, the more likely a similar symbol will confuse. That applies also to the commercial strength of the mark. A mark receives increasing protection if it becomes well known: "The more deeply a plaintiff's mark is embedded in the consumer's mind, the more likely it is that the defendant's mark will conjure up the image of the plaintiff's product instead of that of the junior user."[10]

Questions

1. Is the mark PINEAPPLE confusingly similar to APPLE when used on Apple-compatible computer kits?

Might Be Thought to Sponsor Pineapple Kits

2. Is Microsoft confusingly similar to Megasoft (for information technology service management)?
3. Green M&Ms are aphrodisiacs, an urban legend suggests. A California company begins selling The Green Ones, green candies shaped like M&Ms (but without the M&M logo). Trademark infringement?

4. Is Louis Vuitton confusingly similar to Chewy Vuitton?

5. *Grand Theft Auto* (video game) versus Grand Text Auto (academic blog about computer narrative, poetry, games, and art)—confusingly similar?

Answers

1. APPLE is a well-known and distinctive mark for computers. PINEAPPLE for computer kits—quite different from the word *apple* in most contexts—was held to be confusingly similar on the theory that consumers might well think Pineapple was a name for a new Apple product. By contrast, Mangosoft® did not infringe Microsoft®.

2. Because *micro* and *soft* are such common words in tech, Microsoft is not confusingly similar to Megasoft for technology services. For the same reason, MEGASOFT is a mark for several other businesses, in decreasing order of relatedness: a "Solution for RFID Technology," "payroll outsourcing," "Patient Return Electrode"(?), and "the key to your most comfortable bicycle ride ever."

3. Yes, confusingly similar. The marks are quite different. But the appeal to consumers is premised on the product being M&Ms, a brand that is highly distinctive (well known and quite different from other candy marks). After discussions with Mars, Inc., the infringing company agreed to change the product name to Greenies.[11]

4. Chewy Vuitton summons up thoughts of Louis Vuitton, but not in a way that the two seem to come from the same source. Rather, one is poking fun at the other. A parody will be similar to a mark, but the point is that it is comically different. The song "Barbie Girl," which caustically parodied the apparent worldview of Barbie® the doll, did not infringe. Nor would the following infringe, even if they weren't fictional *Simpsons* businesses: Gritz-Carlton (versus deep-fried Ritz-Carlton), Live Free or Diner (versus New Hampshire's state motto "Live Free or Die"), AD/AD (a Christian AC/DC cover band), or U-Trawl Boat Rentals (versus U-Haul truck rentals).

5. Similar, but amusingly, not confusingly. "Grand Text Auto" valludes to the game but also to the blog's subject matter of

computer-generated literature (on several levels: automatically generated text; automatic writing, as in unconscious authorship; and even AutoCorrect). The blog's discussions of semiotics will immediately set straight any gamer happening upon it. *Grand Thief Auto* for a knockoff video game would display bad faith, a factor supporting infringement. A witty allusion does not.

Proximity in marketplace

If Cadel sold an unauthorized GOOGLE computer, that's confusingly similar to the famous GOOGLE for software and hardware. GOOGLE pomegranates, not so, even if Google has server farms. In other words, context matters.

U.S. Patent   Nov. 15, 1977   Sheet 2 of 7   **4,058,795**

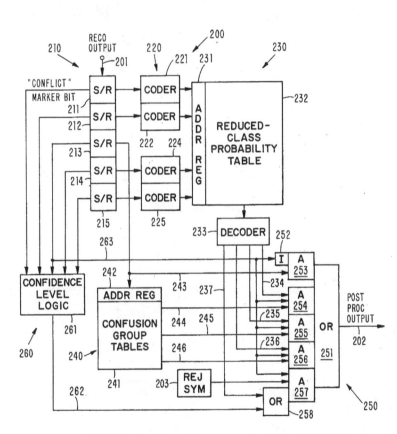

IBM's "Method and Apparatus for Context-Aided Recognition"

DEATH WISH COFFEE®, made with triple the normal caffeine, does not infringe DEATH WISH PIANO MOVERS®. DEATH WISH is quite distinctive, but these are distant product markets. M&M MEAT SHOPS (a.k.a. Les Aliments M&M) for Canadian frozen food stores infringes neither M&M® for candy nor EMINEM® for a rapper.

Are the following confusingly similar?

1. Jumbo jets versus a blog name:

v.

2. Cereal versus gasoline:

v.

3. "Iron Man" (Black Sabbath song) versus *Iron Man* (an *Avenger* movie)
4. Drake University versus Drake (rapper)
5. Mobil uses a flying horse as a trademark. Pegasus Petroleum is named after a mythical flying horse. Trademark infringement?

6. Python.org supports Python, an "interpreted, interactive, object-oriented, extensible programming language."[12] Python literature and tutorials refer often to the Monty Python shows and movies. Trademark infringement?

v.

7. Apple Records starts up in 1967, selling records by the Beatles and others. In 1976, Apple Computer sells computers. Trademark infringement? By 2012, Apple Computer has become just Apple and operates the biggest online music store. Trademark infringement?

v.

8. A teenager operates harrypotterguide.co.uk, a noncommercial fan website offering information about *Harry Potter*. Trademark infringement?

9. *Reader's Digest* versus readersdoglist.com (compilations of lists)

10. *The Straight Dope* readers quickly came up with a list of songs using trademarks not just in the song, but in the title, such as "Kodachrome," "G6," "Aqualung," and many car songs: "Mercedes-Benz," "Hot Rod Lincoln," "Little Red Corvette," "Mercury Blues," "Mustang Sally," "Pink Cadillac." We might add "Rednecks"; "White Socks and Blue Ribbon Beer"; "Radio Free Europe"; "American Girl"; "On the Atchison, Topeka and the Santa Fe"; and "Lullaby of Birdland."[13] Those are commercial uses. Is using someone's trademark to sell records infringement?

11. GREATEST SHOW ON EARTH (circus slogan) versus GREATEST SNOW ON EARTH (Utah marketing slogan)

Answers

1. They sound alike, but these are completely different markets. Therefore, this is not confusingly similar.
2. Using cartoon tigers in utterly different markets would not confuse anyone. But when Esso gas stations started selling competing food products with a cartoon tiger mark, that was a confusingly similar use.
3. No. Nor would either be infringed by the Iron Man triathlon.

4. Ditto.
5. These could be confusingly similar. A flying horse image is easily confused with the name of a flying horse, given the fact that both are used in the oil business.

Had Mobil sued Pegasystems (software company), the result would be different.

Build for Change®

6. These are clear allusions, not confusingly similar.
7. At some point, APPLE COMPUTER did move into the music business, making it confusingly similar to APPLE RECORDS. The two have come to terms.
8. These are in similar markets, but they have distinguishable sources, if it is made clear that the one is a nonsponsored website writing about *Harry Potter*.[14]
9. A gentle parody in the name dispels confusion; it will not appear to be a sponsored website.
10. The uses in the titles are references to marks, not apparent claims to be from the same source. Trademarks may be freely referenced.
11. This is only a reference; skiers will not think Barnum and Bailey runs the lifts.

Buyer's knowledge and attention

 v.

Confusingly Similar?

A sign from *The Simpsons*: "U.S. Air Force base: Not affiliated with USAir." The joke is that no one confuses the two even though they are quite similar. Someone buying a ticket to Palau is unlikely to inadvertently enlist in the military. Someone looking to sign up as a pilot likewise will not buy a ticket to Pittsburgh. When consumers are likely to know quite a bit about the relevant product, and when they are likely to pay a lot of attention to the transaction, confusion is less probable. Conversely (a favorite lawyer's word for switching hands, not to be confused with **CONVERSE**), someone standing in line making an impulse purchase of candy might well fall for a spurious packet of M&Ms even if a few seconds' attention would dispel the confusion.

By the way, USAir changed its name to US Airways in 1997—not to avoid confusion with the U.S. Air Force, but rather because the word *Airways* was associated with larger airlines, whereas *Air* was more likely to be a regional carrier.

Questions

1. Pharmacies package their house brands to look like best-selling brands such as Benadryl and Tylenol and place them right next to each other on the shelf. Trademark infringement? Do these look confusingly similar? Would they at 2:00 a.m.?

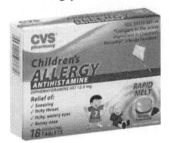 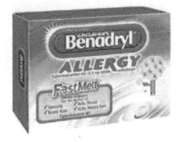

2. Satirist Al Franken published *Lies (And the Lying Liars Who Tell Them): A Fair and Balanced Look at the Right*. The title sarcastically uses the trademark slogan of one of the book's targets, Fox News: Fair & Balanced®. Fox sued. Who won?

3. Facebook logo versus Fedora open source operating system logo: Are they alike? Is there a likelihood of confusion? LOL?

1. Some bleary-eyed cold sufferers or tired parents might get confused, but the brands have not sued. They are not confusingly similar, simply because the average customer knows of the distinction between the pharmacy's brand and the famous brand. The house brand states on its package "Compare to Benadryl," which helps dispel confusion. No lawsuits yet. The pharmacy and famous brands are both competitors (seeking to sell to the same market) and cooperators (the pharmacy needs to stock the famous brand to keep some customers happy, and the famous brand needs shelf space at the pharmacy).

2. Few book buyers would confuse lefty Franken's book to be sponsored by righty Fox News. Fox lost quickly, and, as often happens when trademark owners overplay, the publicity was a boon to Franken. The case may be one of the reasons Franken later became Senator Al Franken. He won by just 312 votes, out of over 2 million. In support of latitude for writers, the Author's Guild submitted a friend-of-the-court brief listing dozens of books using trademarks in the title, including:

   The Greatest Show on Earth by Katherine Leiner
   Cadillac Jukebox by James Lee Burke
   The Lexus and the Olive Tree: Understanding Globalization by Thomas Friedman
   A Mind Is a Terrible Thing to Waste: Coping at the Start of Your Career by Bob Rosner
   The Devil Wears Prada by Lauren Weisberger
   Soy la Avon Lady and Other Stories by Lorraine Lopez
   Breakfast of Champions by Kurt Vonnegut
   The Electric Kool-Aid Acid Test by Tom Wolfe
   Prozac Nation: Young and Depressed in America by Elizabeth Wurtzel
   Napalm & Silly Putty by George Carlin

3. The designs are pretty similar, white F on blue, but different on closer inspection. Fedora's logo has rounded ends and morphs into a faint infinity symbol. The products are both software, but one runs computers and one connects people. The designs pop up in quite different contexts. If Fedora started a social network app, or Facebook started an operating system (both plausible), then there could be some confusion.

Dilution

Famous marks get an additional type of legal protection. The owner of a famous mark may prevent uses that "dilute" the mark. If Cadel used GOOGLE to sell pomegranates, a court might rule that the use must stop because it would reduce the distinctiveness of the famous GOOGLE mark if others use it, even in ways that do not infringe. That would be dilution by blurring. There may also be dilution by tarnishment: using the famous mark in a manner that harms the reputation of the mark, such as using it on unsavory goods. So beware of using famous marks.

Trade Secrets[1]

> *His mind, of man a secret makes,*
> *I meet him with a start,*
> *He carries a circumference*
> *In which I have no part.*
>
> Emily Dickinson (1870-ish)

> *And they encouraged a sort of informational*
> *hygiene, a belief in copying things strictly*
> *and taking great care with information, which*
> *as they understood, is potentially dangerous.*
> *They made data a controlled substance.*
>
> Neal Stephenson, *Snow Crash* (1992)

DuPont closely guarded information on how it made and marketed Kevlar, the lightweight bulletproof material. A former DuPont employee broke his agreement to keep the information secret and, for a price, delivered it to a competitor. The competitor used the information in making and selling its competing product—until it lost a $919,000,000 jury verdict to DuPont for trade-secret misappropriation.[2]

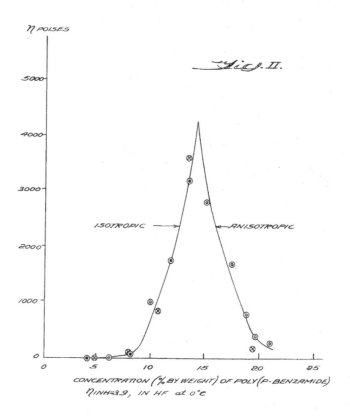

Fig. II.

INVENTOR
STEPHANIE LOUISE KWOLEK

Kevlar Itself—Not a Trade Secret

DuPont's 1974 patent on Kevlar expired long ago. Anyone may make and sell that material freely, even getting help from the description in the expired patent. DuPont's later-developed processes for making Kevlar more efficiently and DuPont's marketing data and strategies were not patented or copyrighted. Anyone is likewise free to do the same things. But the competitor cannot *wrongfully* get that information by bribing an employee to breach his or her duty of confidentiality. The competitor could legally develop the processes itself without infringing DuPont's rights, and likewise could have studied DuPont's product and marketing activity to figure out DuPont's secrets. Trade-secret protection bars only wrongful access to data, such as by bribing, breaking confidentiality agreements, or using industrial espionage.

What information exactly did DuPont's competitor take? That has not been made public. Most court cases are open to the public, as freedom of speech and due process require. But courts often keep trade-secret information sealed because such information is economically valuable precisely because it is kept secret.

Formula[3]

COCA-COLA
SYRUP AND EXTRACT.

For Soda Water and other Carbonated Beverages.

This "INTELLECTUAL BEVERAGE" and TEMPERANCE DRINK contains the valuable TONIC and NERVE STIMULANT properties of the Coca plant and Cola (or Kola) nuts, and makes not only a delicious, exhilarating, refreshing and invigorating Beverage, (dispensed from the soda water fountain or in other carbonated beverages), but a valuable Brain Tonic, and a cure for all nervous affections — SICK HEAD-ACHE, NEURALGIA, HYSTERIA, MELANCHOLY, &c.

The peculiar flavor of COCA-COLA delights every palate; it is dispensed from the soda fountain in same manner as any of the fruit syrups.

J. S. Pemberton;
Chemist,
Sole Proprietor, Atlanta, Ga.

Old Coke Ad

The most famous trade secret (a "famous secret"?) is the recipe for Coca-Cola, still under wraps after more than a century of commercial use. Several other popular foods are likewise made with secret recipes: KFC's secret blend of "eleven spices," McDonald's Big Mac's special sauce, and, until not long ago, Mrs. Fields' cookie recipes. Willy Wonka had well-guarded formulas for candy. A formula for dog food can be a trade secret. Such a formula is easier to keep secret because the dogs cannot tell potential competitors just what is so appealing about the food.

WD-40 lubricant is another well-known (albeit less nutritious) product made with a secret formula. The recipe for Twinkies is likewise secret. Some suggest that consumers might prefer not to know just what gives a Twinkie a shelf life of years. Consumer laws require products like Coca-Cola and Twinkies to disclose ingredients but leave room to protect trade secrets by simply requiring a list of generic ingredients such as "natural flavorings."

Businesses may have their own secret formulas for products from cheese to paint to solvents used in manufacturing. High-tech manufacturing and research may involve the preparation of special substances whose formulas are kept away from competitors. Beer and wine makers often have trade-secret formulas and processes for making their unique products.

Oil drillers using fracking have secret formulas for chemicals used in drilling—a good example of tension between trade-secret protection and government regulation such as environmental protection.

Lots of Secrecy in the Oil Business

Oil roughnecks screwing one piece of drill pipe to another piece by means of a heavy pipe wrench, Kilgore, Texas, Date Created/Published: 1939 Apr. Part of: Farm Security Administration—Office of War Information Photograph Collection (Library of Congress) via totallyfreeimages.com

Formulas for patent medicines and other nostrums were held secret at one time, but today the FDA requires drug sellers to disclose such matters.

Patent Medicines

Competitors may resort to wrongdoing to get valuable formulas. In the 1930s, Procter & Gamble, trying to protect the market for its Ivory Soap, bribed employees of Lever Brothers to turn over experimental cakes of soap and formulas.

An Object of Competitors' Envy

Pattern

Mechanical Creation of a Perspective Image by Albrecht Dürer

John H. Lienhard, Engines of Our Ingenuity, http://www.uh.edu/engines/epi138.htm

Manufacturers often keep confidential the drawings that show how to make important equipment, ranging in size from giant steel-cutting machines to tiny special-purpose electronic circuitry. Trade-secret patterns could include circuit diagrams, machine blueprints, shapes of molds, designs of drilling equipment, or designs of laboratory equipment. The prototype of a new product could be a trade secret.

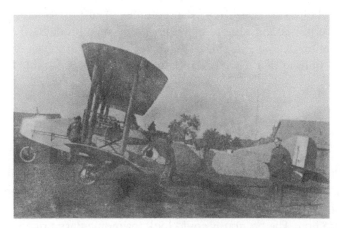

"First Prototype Aircraft of the Avro Type 523 Pike, the Sunbeam-Engine Pusher Version, Completed in May 1916"

Part of collection: E.A. Crome collection of photographs on aviation, via totallyfreeimages.com, National Library of Australia, http://catalogue.nla.gov.au/Record/3722895

Compilation

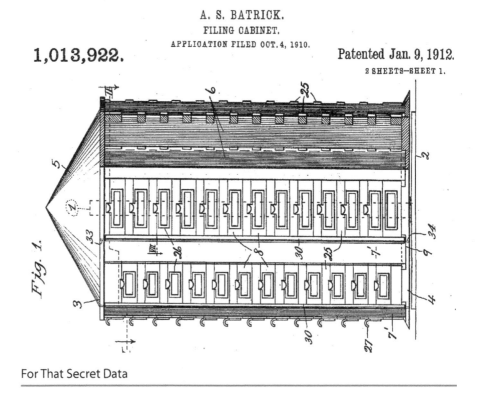

A. S. BATRICK.
FILING CABINET.
APPLICATION FILED OCT. 4, 1910.

1,013,922.

Patented Jan. 9, 1912.
2 SHEETS—SHEET 1.

Fig. 1.

For That Secret Data

Most trade-secret lawsuits allege wrongful taking of business data: client lists, sales records, marketing research, identities of skilled contractors and suppliers. A Home Depot executive, for example, provided confidential pricing information to a supplier as part of a job search.[4] As business moves increasingly online and every interaction is recorded, the volume of valuable data grows exponentially. Amazon has an edge on its competitors because it has gigantic records of what buyers are buying, how they react to various prices, and how their different purchases are related. Locksmiths, jewelers, and clockmakers often keep volumes of information on products for reference in making repairs or in designing new products. Even negative information can be valuable. Thomas Edison tried thousands of materials before finding that carbonized bamboo made for a durable light bulb filament.

Program

Much software is freely disclosed, in its executable form, by selling it to customers. Other programs are more valuable because they are kept secret. Hedge funds use software that detects and exploits opportunities to profit in securities trading. The programs could look for momentary price imbalances that offer arbitrage opportunities for split-second trades—or longer-term strategic investing based on broader market-related data.

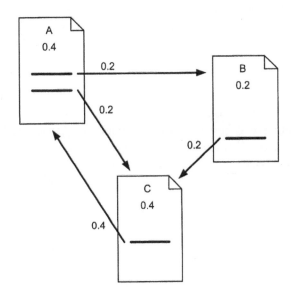

Larry Page, Method for Node Ranking in Linked Database

Google's Page Rank process, which ranks sites based on how other sites link to them, is patented and therefore public knowledge. But Google keeps other aspects of its search methods and rankings secret.

Even where software is sold, trade secrets may be retained. Auto-Tune has changed the music industry by offering pitch correction and manipulation of vocalizing. Auto-Tune's software is distributed to users in a form they can use. But Auto-Tune does not distribute the source code, which would disclose its trade-secret processes.[5]

Device

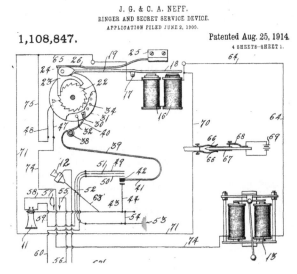

Not a Secret If You Patent It

Many types of devices embody trade secrets. James Bond had all manner of secret weapons designed by Q, although MI6 used extralegal methods to protect its trade secrets. Businesses may develop devices that give a competitive advantage, such as cutting diamonds more quickly, forming perfect lenses, or doing high-speed computing. The device could also do a lower-tech job but in less costly fashion, such as in cutting steel, manufacturing cars, or making food.

Competitors may violate the law to get access to trade secret devices. Engineers from Wyko Tire Technology gained entry to a Goodyear Tire factory on the pretext of doing maintenance work and took photos of a secret roll overply-down device.

Technique

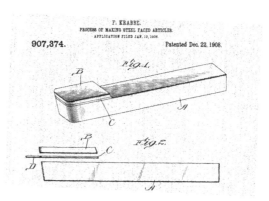

A Patented Process Will Not Be Secret

A common type of trade secret is "know-how": how to use machines, build particular types of factories, operate specialized devices, or test and correct products in manufacturing. Someone who finds a better way to do something often tries to keep it from the competition. Skilled craftsmen may guard special techniques for refined work. Such obfuscation goes beyond factories to the arts. Classical musicians have attempted to conceal techniques, some even refusing attendance to rivals at their concerts. More recently, guitarist Eddie Van Halen would turn his back to the audience or use other tricks to conceal his tapping technique. He patented his "supporting device for musical instruments," shown here, but kept that tapping technique as a trade secret. Magicians naturally conceal the secrets of their tricks, as well.

A supporting device for stringed musical instruments, for example, guitars, banjos, mandolins and the like, is disclosed. The supporting device is constructed and arranged for supporting the musical instrument on the player to permit total freedom of the player's hands to play the instrument in a completely new way, thus allowing the player to create new techniques and sounds previously unknown to any player. The device, when in its operational position, has a plate which rests upon the player's leg leaving both hands free to explore the musical instrument as never before. Because the musical instrument is arranged perpendicular to the player's body, the player has maximum visibility of the instrument's entire playing surface.

Edward L. Van Halen's Patent

Process

Albrecht Dürer, Printer

A slightly faster or cheaper way to make a product can give a huge competitive edge. Trade secrets may arise in any industry: a metallurgy process in high-temperature steel mills, a process to make fiberglass, methods to make chemicals or materials (such as DuPont's methods for making Kevlar). Lexar, a start-up, formed a joint venture with Toshiba. Toshiba later shared Lexar's trade secrets with respect to NAND flash memory with the memory-card maker SanDisk, resulting in a $465 million jury award.[6]

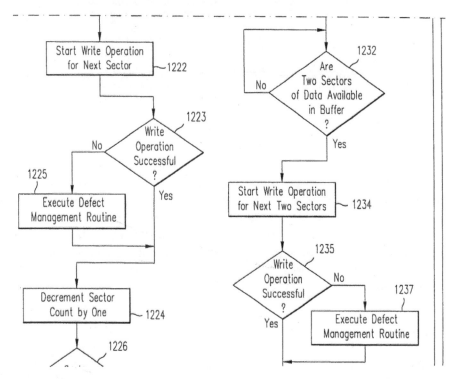

Lexar Also Sued Toshiba for Patent Infringement

The process for making food is often a trade secret, from artisanal cheese makers to bakers. The *New York Times* keeps its method for determining the best-seller lists secret. That gives the *Times* not only an advantage over competitors but makes it more difficult for publishers to manipulate the lists.

Sports are rampant with trade-secret methods. U.S. football teams have strict security around their playbooks. Bodybuilder and former governor of California Arnold Schwarzenegger reportedly kept his best training methods secret, and he misdirected competitors by publishing spurious training tips. He suggested that rather than training once every 24 hours, one could get more sessions in by training every 18 hours. Those who followed the suggestion to stagger training schedules that way might have quickly found their sleep cycle disrupted. Many other athletes keep training and nutritional methods secret, but some are more open: one world-class marathoner, when asked about his special pre-race diet, replied simply, "A bagel and a coffee."

A geologic method to locate oil deposits could be a trade secret. Processes in many different fields can be considered trade secrets.

Other

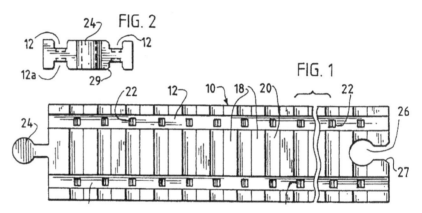

U.S. Patent   Oct. 3, 1995   5,454,513

The Idea for Clickety Clack Track Was a Trade Secret until It Was Patented[7]

Obviously, product ideas and marketing strategy can give a competitive advantage, and thus these could be considered trade secrets as well. Also, negative information, such as disappointing results in testing new materials, can be valuable and hence considered a trade secret.

Most professional baseball teams use Lena Blackburne Rubbing Mud to prepare bats for use. The seller keeps the location of the mud's source as a trade secret. Other vendors and manufacturers also keep sources of supply as a trade secret.

Not trade secrets

Trade-secret law does not provide blanket protection to secrets of all types. In fact, most confidential material would not be a trade secret. The sacred materials of a church or the oral tradition of an indigenous group have value, but not the economic value that would put them in the category of trade secrets. Personal information, such as health records, academic transcripts, and diaries, is confidential, but not trade-secret material. Their value lies in their privacy, not in economic value to potential competitors. Darwin kept his theory of evolution by natural selection as a secret for many decades. A theory with great scientific value but no economic value does not qualify as a trade secret.

economic value to competitors

Even in a business, much information with economic value may not qualify as trade secrets. Lots of the information that employees have from education, training, and experience on the job is necessary for the business. But the information may well be available to others from other sources.

Questions

1. What trade secrets might the following organizations have?
   a. FedEx (delivery services)
   b. Sotheby's (auction house that sells Picassos)
   c. Full Moon (local restaurant)
   d. Walmart (retail stores)
   e. Boeing (airplane manufacturer)
   f. Chicago Bears (a professional football team)
2. Is Coke's "famous trade secret" a contradiction in terms? That is, can it be a secret if it is famous?
3. What trade secrets might a university have?
4. Could computer passwords used in business be considered trade secrets?
5. Could a list of Twitter followers be a trade secret?

Answers

1. We don't know their secrets. But we can speculate.
   a. FedEx: marketing data, processes for handling packages efficiently, computer software
   b. Sotheby's: customer lists, sources of supply
   c. Full Moon: recipes
   d. Walmart: supply chain software, price and marketing data
   e. Boeing: manufacturing techniques and processes
   f. Chicago Bears: playbooks
2. The formula itself is still secret, although it is well-known that there is a formula.
3. Universities have protected such information as lists of donors and marketing information, but they usually take the opposite approach of encouraging publication of valuable information.
4. Passwords as trade secrets? Some would reason that passwords are merely private symbols, of no commercial use to competitors. Others figure that the only value of passwords is that others do not know them, so that would indicate economic value from secrecy.
5. A list of followers can have value from being not known to others, as with a customer list.

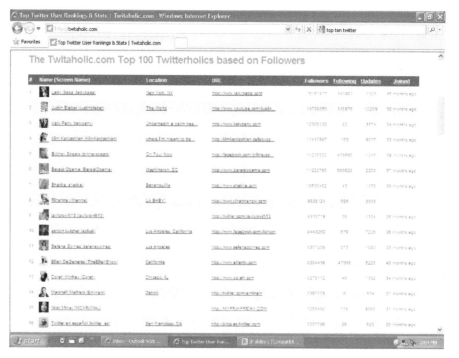

A List of Twitter Followers Has Value

Security measures

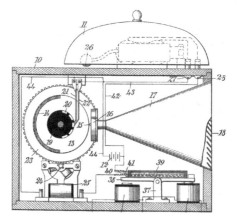

No. 873,638.

PATENTED DEC. 10, 1907.

C. VAN BERGH.

BURGLAR ALARM.

APPLICATION FILED JUNE 6, 1907.

A Security Measure

A proud factory owner gives company tours. An entrepreneurial tourist sketches the designs of the factory's unique high-tech bean-processing machines and soon sets up her own business. No trade-secret stealing here because there was no trade secret. Information is a trade secret only if the owner uses security measures. Allowing random members of the public to view the machines is nice, but it indicates there is no security system. To have legal protection, the factory should exclude unnecessary visitors and have others sign nondisclosure agreements.

Perfect security is not required. One could ensure secrecy by destroying all copies of the information, *Mission: Impossible*–style. However, to get value from a trade secret, one must tell it to people: employees, joint venture partners, investors, and designers of custom-made devices.

The best approach is to put a security program in place.

Identify trade secrets

The first step in security is for the business to canvass its operations to identify what trade secrets it has. Depending on the business, this procedure could range from thinking things through with the founder of a start-up to making an audit of a chemical company with several divisions. Beware of the trap of treating all internal information as "secret." The increased security would interfere with actually getting things done, and treating trivia as treasures will lead to employees disregarding security as foolish.

Provide nondisclosure agreements

The trade-secret information should be shared only with those who need it and are trusted. Furthermore, it should be shared only after the recipient has signed an agreement not to use it or disclose it. Nondisclosure agreements should be signed by employees, vendors, and joint venturers.

Keep track of copies of trade-secret information

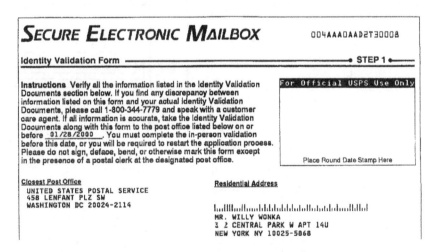

Methods and Systems for Proofing Identities Using Certificate Authority

When information is truly a trade secret, its location should be fixed. Documents should be stamped with CONFIDENTIAL, EAT AFTER READING, or other inspiring phrases. Electronic files can also bear similar legends, not to mention be protected with passwords and tracked. Information may take many different forms: software, customer lists, drawings, formulas, videos—and security measures can be tailored to each.

Limit access to trade-secret information

Limiting access to trade-secret information helps keep track of copies, reduces the number of people who might abuse it, and encourages those who do have access to be careful—because they know that leaks of the information could come only from a limited number of sources. The limits also emphasize the confidential nature of the information. Only those who need the information should have access to it.

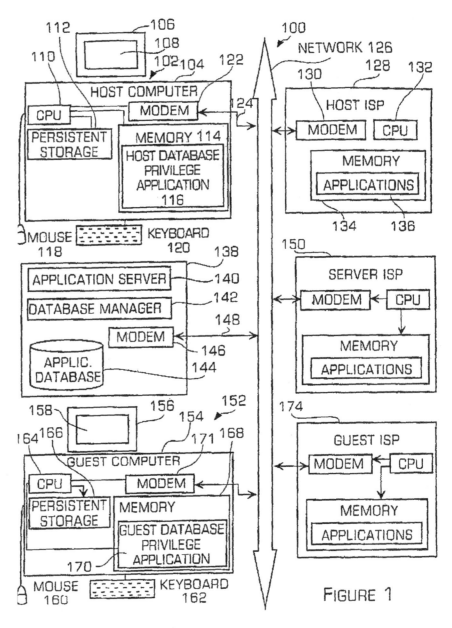

FIGURE 1

This Agile Software Development Group Patent Description Reads, "This Is Particularly Useful in Complex Business Relationships between a Host and a Plurality of Users, Both of Which May be Sensitive about Their Trade Secrets"

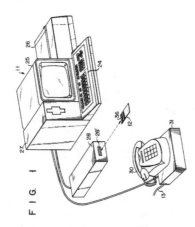

Patent for a Data Processing Device with High Security for Stored Programs, by Toshiba

Finally . . .

F. H. AULD.
WORKMAN'S IDENTIFICATION BADGE.
APPLICATION FILED APR. 15, 1918.

1,298,273.   Patented Mar. 25, 1919.

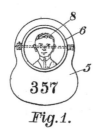

Fig.1.

Old School

Educate employees about the need for secrecy

Things regarding security that seem obvious to the business owner may not occur to the employee concerned with tasks at hand. Employee training can standardize procedures and encourage employees to get with the program.

June 5, 1951   O. F. SCOTT ET AL   2,555,422
                         LIE DETECTOR
                      Filed June 19, 1947

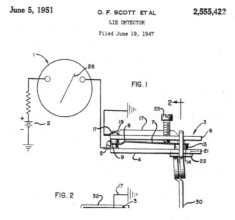

Not Necessary for Exit Interview

Use trade-secret information in separate location (such as a secure area in a factory)

In some defense contractor premises, a light flashes whenever a visitor or newish employee without a high-security clearance is present. When the light goes on, office doors close and cubicles hush. In some settings, the visitor actually carries the light with them.

U.S. Patent   Dec. 30, 1986   **4,633,231**

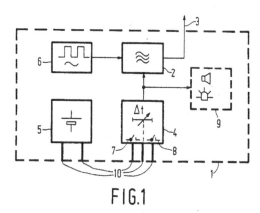

FIG.1

Monitoring Arrangement for Visitors

Lock up trade-secret information (physically, in a safe, or logically, in encryption)

After starting the security program, make periodic checks to make sure people are with the program. Security can be annoying and obstructive. Soon people will put sticky notes with passwords on their computers or take information on vacation (to view while sitting next to a stranger on a train). People also have good reasons to share information—exchanging tips with others in the profession or publishing papers to promote scientific, medical, or professional achievement. So people will need both encouragement and reasons why some information should be kept confidential. Beware, however, of marking too much information as confidential, which undermines security for the genuine secrets and introduces all kinds of inefficiencies. For example, software developers may get assignments without being told what data will be coming in and going out. The costs of security should be weighed against the advantages of keeping the information secret.

Using the Internet creates chances of making trade secrets more accessible from the outside. IBM reportedly prohibits employee use of Apple's Siri, and other companies limit use of Google's services for fear of employees indirectly revealing important information that way.[8] Paranoid? Well, as your author googled some of the cases in this chapter, ads popped up for spy cameras, aerial surveillance, and employee monitoring. Search engine activity can convey information.

> Internet **Employee** Monitor
> Internet.usage.control.gfi.com/
> Now with Search Engine **Monitoring**!
> Monitor **Employees** Search. Try Free.

Use targeted advertising to check security

Businesses should keep tabs on their competition. They do that anyway, for marketing and competition. They can observe what the competition is up to, and whether it indicates that confidential information has been disclosed. Some businesses include a dummy address (a fictional person at a real address) or two in their customer list. That way they know that if a solicitation arrives addressed to the fictional person, the list must have made its way to the sender.

Questions

1. Coca-Cola has put its secret formula in a locked vault—on display to the public in *The World of Coca-Cola* exhibit in Atlanta.[9] Can a formula on a plinth be a trade secret?
2. Would a typical college student have trade secrets?

Answers

1. Perfect security is not required. Coca-Cola can balance the costs of security against the advantages of secrecy. The formula is locked in a vault under the watchful eyes of caffeinated guards, even though its location is known to the public. The great publicity for Coca-Cola must also enter the equation. Others flaunt their secrets: KFC nurtures the folklore around its secret recipe, such as the story that no one person knows the complete formula and that those in the know may not travel together, for fear of catastrophe. One might wonder whether the formula is really so crucial for KFC's success, as opposed to the company's having good marketing and efficient operations. As a savvy observer put it, "I'm willing to bet if someone released the recipe to the KFC chicken, people would say, 'Oh, well, that's not much different from my grandmother's.'"[10] But if recipes have some value from secrecy, and there are security measures, trade-secret protections apply.
2. Some entrepreneurial students have their own businesses, so they might well have customer lists, proprietary software, secret product plans, even formulas developed during a chemistry lab. But a typical college student would not have trade secrets. A student would likely have all kinds of confidential information (ranging from secrets confided by friends to their true feelings about the future), but those would not be trade secrets unless they had economic value

from being secret. Some students might have info like that (a product idea, some nifty software they wrote, some science that could yield money), but even that would not be a trade secret unless they actively took security measures.

indefinate term

Trade secrets last until disclosure

A trade secret lasts until it becomes public information. A secret product idea is not secret once the product is publicly marketed. Valuable information may also be published in scientific papers. Competitors may reverse-engineer a product or process. Trade secrets are often accidentally revealed (e.g., hitting "reply all"). Sometimes trade secrets are revealed as part of seeking government approval for a product, like drugs, or in court cases. Apple has revealed confidential information as part of enforcing its iPhone and iPad patents.[11]

Question

Magician Horace Goldin patented his ingenious illusion for sawing a person in half. He shared it with other magicians under strict nondisclosure agreements. Someone nevertheless shared it. Trade-secret misappropriation?

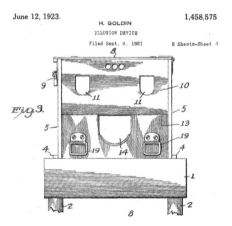

Answer

This was no trade secret, so there was no trade-secret misappropriation (but it may have been breach of contract). The patent publicly disclosed the method of the illusion, so it was no longer a trade secret.[12] After losing the case, Goldin did not patent his other tricks.

Misappropriation (wrongful access, use, or disclosure)

Sega did not disclose the source code to its blockbuster Genesis video game system. Accolade painstakingly reverse-engineered a Sega game, meaning

Accolade took the executable code, studied it in detail, and figured out the necessary codes. Accolade may have figured out Sega trade secrets, but it did not break the law. Trade-secret law protects only against wrongful access—bribery, extortion, breaching nondisclosure agreements, burglary, espionage, and the like. If Accolade had bought trade secrets from a Sega employee or had broken into Sega's offices or hacked into Sega's network to get the code, that would have been illegal. Taking apart a product you've bought and figuring out how it works is not wrongful access to information. It is research—exactly the sort of thing trade-secret law exists to encourage.

Likewise, it is legal to figure out others' trade secrets by talking to their customers, analyzing market data, or perusing public information such as reports of public agencies. Some market their services for such analysis. A patent application on "Method and apparatus for the discovery and reporting of trade secrets" included "the steps of collecting trade secret information from a plurality of persons of the organization, analyzing the trade secret information using mathematical and logical formulae to identify a plurality of trade secrets of the organization, and generating a report regarding the identified trade secrets of the organization."[13]

But unfair means are wrongful, even if no bribery or burglary is involved. DuPont built a plant in Beaumont, Texas, in the 1960s to use a trade-secret process to make chemicals. A competitor hired some locals to fly over the construction site and take photographs. The competitor argued that it was like reverse-engineering, simply taking photos from public airspace. The court ruled that the conduct fell below accepted standards of business behavior. Otherwise, parties would have to overinvest in security measures, such as building a tent over a construction site to keep out flying eyes.

Trade Secrets at the Intersection of DuPont Road and DuPont Road?

Procter & Gamble conceded trade-secret violations after spying on Unilever, a competitor in the consumer goods market, including searching through Unilever's trash. Searching through trash may be legal, admirable even, for recycling purposes, but doing so looking for trade secrets falls below standards of commercial ethics, as in the DuPont case. High-value trade secrets should be shredded before disposal, but not every reference to marketing strategy or new products can be tracked and destroyed.

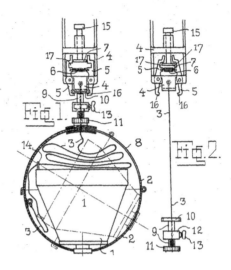

No. 894,348.   PATENTED JULY 28, 1908.

G. B. SEELE.

APPARATUS FOR TAKING PHOTOGRAPHIC PICTURES FROM A HEIGHT.

APPLICATION FILED AUG. 29, 1907.

Aerial Photography Patent

Questions

1. Chemists at Artisanal Cola analyze samples of Coca-Cola and figure out the secret formula. Artisanal Cola uses it to make and sell Secret Sauce Pop. Trade-secret infringement? What if Artisanal had hired an ex-employee of Coca-Cola and she disclosed the formula, contrary to her signed nondisclosure agreement?
2. How about using Google Earth to study a competitor's chemical plant and figure out its process for isolating tungsten?

How Far Can You Go?

Retronaut.com

3. Would it be trade-secret misappropriation to follow FedEx's trucks around to figure out their efficient delivery process? How about planting a GPS device on a FedEx truck to do it remotely?

Answers

1. Figuring out trade secrets by reverse-engineering is not wrongful. Paying an ex-employee to break her promise not to tell the formula is wrongful, as would be stealing a competitor's secret device in order to study it.
2. Aerial photography of private property may be wrongful, but use of a commonly available software application would seem above the murky line. Google's many services are making information easier to get, meaning security measures must adjust.

Using Pigeon Spy Cameras Would Be Misappropriation

Retronaut.com

3. Following trucks on public roads would seem not wrongful. Planting a GPS device may be more environmentally friendly, but definitely below the line.

Trade Secrets and Employees

> [Dilbert® briefs colleagues on security procedures for valuable trade secrets:]
> *How much would our competitors pay for our secrets?*
> *Oh, I dunno . . . maybe several times your annual salary.*
> [Employees exchange evil grins]
>
> Scott Adams, *Your Accomplishments Are Suspiciously Difficult to Verify*

Borland alleged that Symantec hired a Borland employee to get trade secrets regarding development and marketing of security software (ironically enough). eBay claimed Google stole its trade secrets in order to develop Google Wallet by hiring two eBay employees. Boeing claimed Lockheed Martin hired its employees to get trade secrets in rocket making. A former Intel employee took information valued in the millions to a new employer, and Intel rival, Advanced Micro Devices. Samsung alleged that employees sold its competitor LG trade secrets on organic light-emitting diode televisions. An employee of Coca-Cola was arrested for attempting to sell Pepsi-Cola product information along with a free sample of a secret beverage under development.[1]

Businesses Depend on Employees

*"Women workers predominate in English factory scene during
the Great War (1914–1918), aka WWI." 1916 (Google Life Archive)*

Employers contend with the classic problem from economics, the principal/agent problem. People rely on others to do lots of things because of specialization of labor (a surgeon will likely do a better job on that heart transplant than the patient himself would) and efficiency (you could personally deliver that letter to Aunt Mame in California, but it is easier to pay the USPS® 45 cents to do it). But the agent has different incentives than the principal. The stockbroker may embezzle the customer's funds or invest them carelessly, mail carriers have been known to toss the odd bundle of mail into the Chicago River, surgeons' minds may wander. Employees may be the biggest asset of a business—or they may divert the business's assets for their own use. *Dilbert*'s Wally: "This week I sold company secrets, did some insider trading, and took kickbacks from vendors."

Trade Secrets on Their Way Home for the Night

Library of Congress Flickr stream

Most trade secret cases involve employees: employees leaving to work for a competitor or to start their own business, or, less often, employees simply selling information to a competitor. Willy Wonka responded to theft of his candy-making secrets by shifting his workforce to Oompa-Loompas, who lived in the candy factory.[2] The city of Venice prohibited its glassmakers from leaving the city (as other cities had controlled their craftspeople and holders of secret arts).[3] A nonfictional, modern employer does not have those options. Trade-secret law by itself provides only limited protection to employers because misappropriation can be difficult to prove and the company may be closing the barn door after the horse has bolted. Consequently, employers often seek to control information by means of employee contracts.

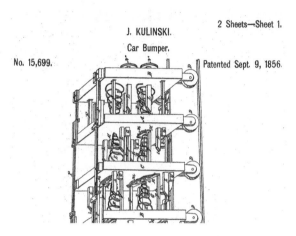

J. KULINSKI.
Car Bumper.

No. 15,699.

2 Sheets—Sheet 1.

Patented Sept. 9, 1856.

Contracts Provide Protection

Like driver's license tests, employment contracts are written with past disasters in mind. A driver's test tracks accident reports: "If you miss your exit off the highway, should you drive in reverse back to the exit? On long drives, should you conserve space by following the car in front as closely as possible? If you see two headlights coming toward you at night, should you try to drive your motorcycle between them?" All this is written in the hope of educating future drivers and avoiding common causes of accidents.

Trade Secrets on Board: Drive Carefully

Employers likewise learn from experience and thus try to avoid losing trade secrets because employees blabbed, fibbed, defected, conspired, or bumbled. Following is a discussion of some common clauses in employment contracts. Courts will not always enforce such clauses. In general, they will not be enforced if they are too strict. Whether clauses are effective may vary according to the law of the state involved. California law, for example, allows employers less leeway in restricting employees than Massachusetts law. Some say that legal differences are one reason California's Silicon Valley has led Massachusetts's tech companies in commercial innovation. California employees may be more mobile due to less enforcement of restrictive clauses. Expertise and information circulates faster, leading to a more dynamic environment.[4]

Employment contract clauses

Nondisclosure clauses

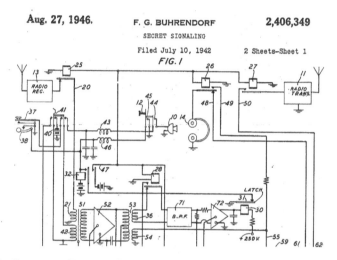

Aug. 27, 1946.   F. G. BUHRENDORF   2,406,349

SECRET SIGNALING

Filed July 10, 1942   2 Sheets-Sheet 1

FIG. 1

A Bell Labs Patent on Keeping Secrets

In a typical employment contract, the employee will agree not to disclose confidential information to anyone without permission from the employer. This imposes a legal duty of confidentiality. It emphasizes to the employee the importance of keeping information confidential. If an employee has not signed a nondisclosure agreement, she may be perfectly free to use or share information that she learns on the job.

> The vanity of being known to be trusted with a secret is generally one of the chief motives to disclose it.
>
> Samuel Johnson (1750)

The nondisclosure agreement and policy, like information security generally, is not aimed just at dishonest employees. The greater risk is the honest but momentarily thoughtless employee who gives out the information for innocent reasons: to impress others; to get customers or glory for the employer; to share information with others working in the same area; to publish in a scientific or industry journal.

Questions

1. Connor leaves his job at Intergalactic Rockets. His contract provides that he will not disclose or use the trade secrets of Intergalactic

Rockets for five years. Is that a clause that a court will enforce?

2. Nikita works for five years as a shoemaker for Arch Leather, learning many techniques known only at Arch from master shoe-makers there. Nikita leaves to work for a competitor. If Nikita uses Arch's techniques, would that infringe trade secrets?

Answers

1. Five years is a long time. But the clause leaves him free to develop his own information, prohibiting only the use of information from his employment at Intergalactic Rockets. That would seem OK.
2. Unless Nikita has signed an agreement not to use the information, or unless there was a clear understanding (not easy to show in court), then Nikita is free to use whatever information she learned at Arch.

Noncompete clauses

Competition Drives People

An employee will often agree not to work for competing firms in the same industry for a period of time after leaving employment. That is simpler for the employer than trying to determine if the employee is wrongfully using information.

Question

Connor's contract states that he will not work for any company in the transportation industry for five years. Too strict?

Answer

That clause goes farther than necessary to protect the employer. First, it is too broad, going well beyond rocketry to prohibit him from driving a taxi. Second, to keep him out of his chosen line of work for five years is too long, even to protect against his using information. These clauses get litigated all the time, and judges limit them strictly.

A savvy company could achieve the same result by employing him for five years as a consultant after he leaves, although that entails compensating him for staying out of the rocket game.

Invention disclosure and assignment clauses

An employee will likely agree to disclose inventions, copyrighted works, ideas, trade secrets, or other "intellectual property" developed during employment and agree the rights will belong to the employer.

Archimedes, brainstorming in the tub, realized he could determine an immersed object's density by the change in the water level. Excited, he reportedly ran *au naturel* through the streets of Syracuse yelling "Eureka!" An employment contract will encourage more discreet disclosure of inventions in case the employer wishes to keep them secret. The invention disclosure clause will also have a confidentiality provision.

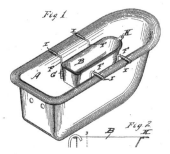

G. C. STRACHAN.
BATHTUB.
APPLICATION FILED JAN. 8, 1919.

1,348,250.

Patented Aug. 3, 1920.

Archimedes's Brainstorming Place

Document retention clause

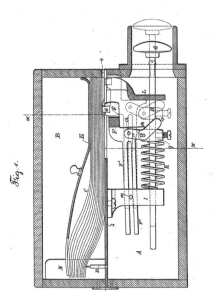

(No Model.)

E. L. MOODIE.
TOILET PAPER SAFE.

2 Sheets—Sheet 1.

No. 416,340

Patented Dec. 3, 1889.

Documents Should Be Safeguarded

Trade-secret information is less useful if the papers are lost or become public. Employees should be encouraged to keep important documents in an orderly fashion. As with all security, balance is important. Only genuinely important papers should be under strict control. If every letter or memo is deemed PROPRIETARY, then pretty soon both important and trivial documents will be floating around unsupervised.

Agreement to follow security procedures

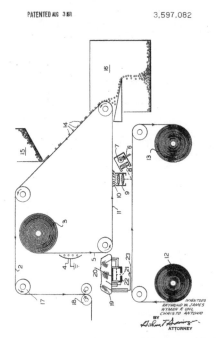

PATENTED AUG 3 1971

3,597,082

"Uncopyable Photochromic Paper"

The employment contract will require the employee to follow security procedures. A careful employer will take time to educate employees on the importance of the procedures and how to follow them. The employer should also monitor how employees actually behave. The best procedure on paper is of no use if employees gradually stray, usually because the security measures are too burdensome or seem pointless.

Trailer clauses

A few months after Tezla leaves her job at Denizen Electric, she markets her latest invention, the battery-charged battery charger.[5] Denizen Electric suspects she developed it while on the payroll there but cannot prove it. A trailer clause seeks to avoid the issue, giving the employer rights to inventions or creative works developed for a period of time after the employee leaves.

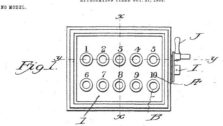

No. 775,348.   PATENTED NOV. 22, 1904.

T. DOW.

MACHINE FOR KEEPING THE TIME OF EMPLOYEES.

APPLICATION FILED OCT. 31, 1903.

NO MODEL.

Fig.1

Time Is Money—and Opportunity

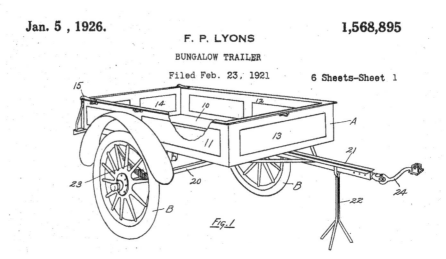

Jan. 5 , 1926.   1,568,895

F. P. LYONS

BUNGALOW TRAILER

Filed Feb. 23, 1921   6 Sheets—Sheet 1

Fig.1

A Long Trailer Clause May Not Be Valid

Third-party contact clause

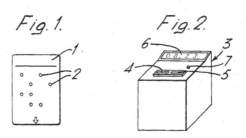

Dec. 31, 1968   MITITAKA YAMAMOTO ET AL   3,419,881

EMPLOYEE CARD SENSING AND RECORDING DEVICE

Filed April 4, 1967   Sheet 1 of 2

Fig. 1.   Fig. 2.

Security May Require Monitoring

A clause in the contract may prohibit contact with certain competitors or require the employee to inform the employer of such contacts. These clauses, if used at all, should be carefully limited. Employees of every sort—software developers, chemists, human resourcers—learn from their peers. They learn not just general know-how and the latest technological improvements in the field, but also comparative information about how to run businesses, for better or worse. Cutting off employees from such information is costly. Rather, only where specific employees and specific risky outsiders are identified should such limits be imposed. It is more important and more effective to encourage employees not to disclose trade secrets to anyone than to have too many limits on access to information.

Nonsolicitation clause

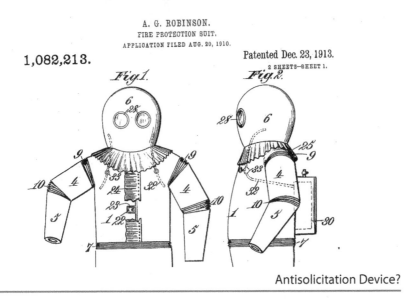

Antisolicitation Device?

An employer cannot prevent employees from leaving, although they can be offered good reasons to stay. An employer can get an agreement (best obtained at the time of hiring, as with the other clauses) that the employee will not try to get other employees to join a competitor.

Nondisparagement clause

A cousin to the nondisclosure clause, the nondisparagement clause requires the employee to agree not to bad-mouth the employer, especially after going to work for a competitor. Enforcing such clauses is tricky, for freedom-of-speech reasons. But the clause may be a speed bump to encourage an employee not to spread spurious rumors about a former employer.

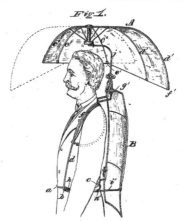

R. RAY.
Sunshade.

No. 229,912.   Patented July 13, 1880.

Fig 1.

Some Contract Clauses Try to Avoid the Light

Debriefing clause

The employer is well advised to sit down and talk to employees when they leave the company in order to remind them of their agreement and give them a chance to disclose information.

"No cold-calling" agreement

Beyond contracts with their employees, companies may take other measures to prevent employees from taking information to competitiors. Some of these measures may be legally hazardous. A class action lawsuit alleged that notable high-tech companies attempted to keep employees in place by agreeing not to recruit each other's employees, creating, in effect, a "no cold-calling" cartel. As with cross-licensing patents, this would be a truce of sorts among competitors. If there was such a deal (the case is pending), the agreement would be illegal, like price fixing. It would short-circuit the market for employment and so be against antitrust laws. The good news is that such agreements between competitors are notoriously leaky for lack of mutual trust.

Hampers and hammocks

Google employees hardly need leave the Googleplex, any more than Oompa-Loompas need leave Wonka's Chocolate Factory.

Google provides its employees with hammocks, grub, Zumba classes, and time to work on side projects. Not to mention stock options.

Googlers Waiting for Swedish Massage

No. 635,261.

J. C. LASSITER.
HAMMOCK AND ELEVATOR.
(Application filed June 4, 1898.)

Patented Oct. 17, 1899.

(No Model.)

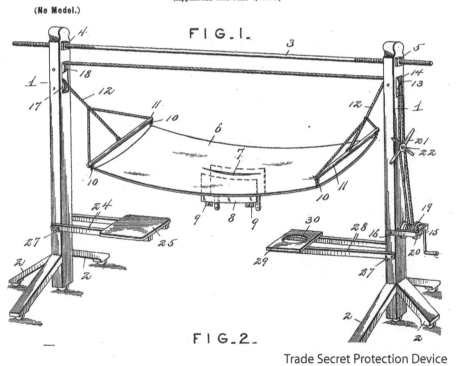

Trade Secret Protection Device

Software Developers Taking a Break from Coding

Basketball games at annual field day at the FSA (Farm Security Administration) farmworkers community, Yuma, Arizona, Date Created/Published: 1942 Mar. Part of: Farm Security Administration—Office of War Information Photograph Collection (Library of Congress)

Employees are the biggest hazard to trade secrets. Unhappy employees are more likely to defect or to sell trade secrets. Harried employees forget security procedures. Unfairly treated employees are more likely to cheat their employer. There are many other reasons for Google to coddle its crew: productivity, collegiality, teamwork, and creativity may all benefit from treating employees well.

Scrooge Coddling Bob Cratchit

Providing food and hammocks may incline employees to spend more time on work. Guarding trade secrets is an additional benefit. An employer need not provide masseuses, mochas, and mints. The key is respect and fair treatment. And coffee.

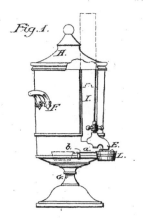

CORNELIA HITCHCOCK.

Coffee-Urns.

Reissued

Coffee Helps Keep the Flock from Straying

International Intellectual Property

Would it not be incompatible with all freedom, if an American's mind must be entirely starved and enslaved in the barren regions of fruitless vacuity, because he doth not wallow in immense riches equal to some British lords?

Robert Bell (circa 1760),[1] relying on Blackstone's *Commentaries on the Laws of England* (1769) (which itself was reprinted by Americans)

Led Zeppelin® had a valid, indeed famous, mark in most countries. But the trademark (varemærke) was deemed rotten in the state of Denmark. Not as a deceptive mark: Danes, and not just Niels Bohr, know zeppelins made of lead cannot fly. Rather, Countess Frau Eva von Zeppelin guarded the name of her grandfather, the inventor of the dirigible. In Denmark, Led Zeppelin performed as "The Nobs."

First Zeppelin Flight (1900)

First Zeppelin Album (1969)

Law is territorial. Each country has its own laws on copyright, patent (pronounced with short *a* in the United States, long *a* in the United Kingdom), trademark ("trade mark" in the United Kingdom), and trade secret. The law has its local flavor. Trade secrets may be American *know-how* or French *savoir-faire*: the connotations are quite different.

American brands have met local resistance in other countries. In the United States, hamburger vendor McDonald's has used trademark law to prevent other businesses from using the Mc prefix—even McDental. Under Irish and Scottish trademark law, not so much. Budweiser, "King of Beers" in the United States,[2] could not prevent Czech brewers of Ceske Budejovice from using their German name, Budweiser Budvar.[3]

Not King of the World

Global disputes become splintered. If Apple accuses Samsung of infringing iPhone and iPad patents, Apple may sue in several different countries—and may get different results. In Cool Brittania, a judge stated that the Samsung device was not "cool" enough to infringe iPad's design patents. In the United States, a jury held that Samsung's phone did infringe the iPhone design and utility patents, to the tune of $1 billion. No other country rests so much faith in democratic procedures as to allow a dozen random citizens to decide billion-dollar high-technology disputes. Meanwhile, judges in South Korea held that Apple and Samsung had violated each other's phone patents, which could theoretically have led to both being taken off the South Korean market. That would be good news for the Windows phone. The *Apple v. Samsung*[4] World Tour of Telephone Litigation continued like a Led Zeppelin itinerary in Germany, Japan, Australia, and beyond. No lawsuits in Antarctica yet, where cell-phone reception leaves something to be desired.

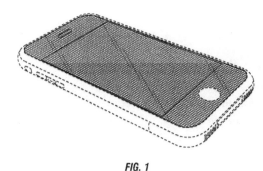

FIG. 1

Apple U.S. Design Patent for "Electronic Device"

Secrecy

Information policy once meant guarding secrets from other countries. The Hittites protected the secrets of forging iron, making their swords stronger than their Bronze Age rivals. Makers of porcelain in Jingdezhen took considerable efforts to keep that process secret, but Europeans infiltrated the manufactories and sent back information and clay samples. Venice in the 15th century prohibited its glassmakers from leaving the city in order to protect the secrets of making stained glass. Brazil was unable to stop others from getting rubber trees, which were then grown even more successfully elsewhere.

Boston owes much to industrial espionage. After breaking free from Great Britain, Americans turned quickly to stealing her trade secrets. Francis Cabot Lowell toured England with his family, socially engineered his way past security into textile mills, and came back with the design for the industry's crown jewel, the Cartwright loom. Textile manufacturers guarded their secrets from each other but underestimated the American bumpkin. English law forbade the export of spinning and weaving machinery and the emigration of textile workers.[5] Lowell relied on his memory, his belongings having been searched for industrial designs before his return to the United States. Soon a booming textile industry arose in a new city: Lowell, Massachusetts. Lowell's family prospered: "And this is good old Boston, The home of the bean and the cod. Where the Lowells talk only to Cabots. And the Cabots talk only to God."[6]

As *Apple v. Samsung* illustrates, nations have gradually, often grudgingly, given rights to foreign authors and inventors. But even after patents became available to foreigners, there were reasons to maintain secrets from foreign competition.

"But many of the best innovators in what was the high technology of the day came from some of the smallest countries in Europe, and these nations did not have patent laws.". . . Swiss inventors tended to concentrate their efforts in watch making and specialized steel making for scientific and optical instruments. Their innovations were exceedingly difficult to reverse-engineer and thus were successfully guarded as trade secrets. "There were competitions in England to reproduce some Swiss innovations in steel, . . . But the English just couldn't figure out how to do it. The Swiss would have been silly to patent these innovations."[7]

National pride

> The people of Spain think Cervantes
> Equal to half-a-dozen Dantes;
> An opinion resented most bitterly
> By the people of Italy.

Edmund Clerihew Bentley

Wilbur and Orville Wright invented the flying machine. The "1903 Wright Flyer, the world's first successful airplane" occupies a place of pride in the Smithsonian's National Air and Space Museum, in the display titled "The Wright Brothers & the Invention of the Aerial Age."[8] The "historic craft that ushered in the age of flight" appears along with many other Wright brothers' items. The Wright brothers' patents, and litigation to preserve their patents, often appear in discussions of American patent law and technological development because those patents are exemplars of "pioneer" patents that open up a new area of technology.[9] The American story of aircraft development runs from the Wright brothers to Boeing and the Space Shuttle.[10]

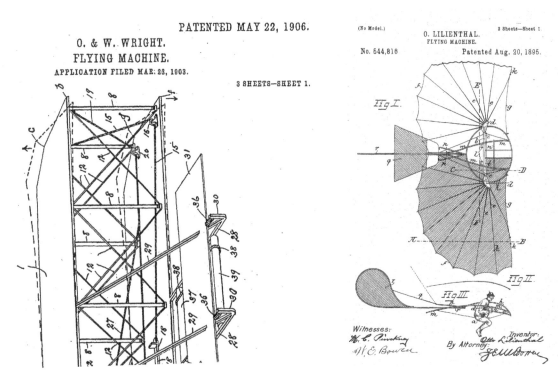

The German Museum in Munich tells a different story: Otto Lilienthal invented the flying machine. The museum's prized exhibition hall for aerospace and astronautics features "50 original aircraft exhibits ranging from Lilienthal to Airbus." A different view of what constitutes an aircraft (a heavier-than-air glider or a powered craft) can support a different story of where it was first developed. An Italian might nominate Leonardo da Vinci, who invented the idea of flying machines even though he did not build any.

Da Vinci, Study (1490)

A more local geographic attribution competition involves the Wright brothers themselves. The brothers worked on their planes in Dayton, Ohio, and flew them in Kitty Hawk, North Carolina. The license plates of Ohio read "Birthplace of Aviation," while North Carolina's proclaim "First in Flight." Ohio changed its slogan slightly to "Birthplace of Aviation Pioneers" so as to include the many astronauts from Ohio, including John Glenn and Neil Armstrong. Stephen Colbert observed, "Twenty-two astronauts were born in Ohio. What is it about your state that makes people want to flee the earth?"

Many important inventions are claimed by multiple nations. Gutenberg is synonymous with the printing press, but in Holland credit often goes to Laurens Janszoon Coster. In the United States, the inventor of the light bulb is considered to be Thomas Edison (patent 1879). In the United Kingdom, one is likely to name Joseph Swan, who patented a less practical forerunner in 1849.

Norwegians, during the World War II German occupation, wore the humble paperclip, which they mistakenly believed to be a Norwegian invention, as a secret (for a while) symbol of resistance to the Nazis.[11] The United States had to use diplomatic means to settle relations with Australia after reports that the United States would ban import of Vegemite® (as in "He just smiled and gave me a Vegemite sandwich" in the song "Down Under"). The inventor of the Swiss Army knife was spurred by learning the Swiss Army sported knives made in Germany. A patent application by McDonald's on a process for making a sandwich caused indignation in the Earl of Sandwich's homeland.[12]

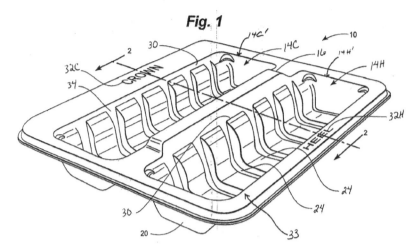

Fig. 1

Ronald McDonald versus the Earl of Sandwich

War concentrates the collective minds of a nation's inventors. The United States and United Kingdom could shoot down enemy planes at night during World War II after developing radar. They inventively concealed that invention. Pilots spread the false story that they ate carrots to improve night vision. The spurious tale spread so well that it is still widely believed. Another innovation was intended to help those pilots if they were shot down. Because pilots might be taken prisoner, British engineers developed a special pencil to hold hidden maps for use upon escaping.[13]

Some attribute creativity to cultural influences:

> Like the fella says, in Italy for 30 years under the Borgias they had warfare, terror, murder, and bloodshed, but they produced Michelangelo, Leonardo da Vinci, and the Renaissance. In Switzerland they had brotherly love—they had 500 years of democracy and peace, and what did that produce? The cuckoo clock.[14]

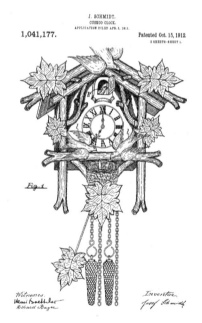

Cuckoo Clock from Josef Schmidt of Villingen, Germany

Or,

NASA's astronauts discovered that ballpoint pens don't work in zero gravity. So NASA spent twelve million dollars and more than a decade developing a pen that writes under any condition, on almost every surface. The Russians used a pencil.[15]

When the U.S. Supreme Court listed patented inventions that "push back the frontiers of chemistry, physics, and the like," it listed patents that were all "Made in America" and patented in the USPTO.[16]

The Marconi Wireless Telegraph Company of America may have wondered if such national feelings crept into patent jurisprudence. In 1905 the USPTO issued patents on radio technology to Italian inventor Guglielmo Marconi. In 1909 Marconi won the Nobel Prize in Physics in recognition of his "contributions to the development of wireless telegraphy." In 1943, after decades of dispute, the Supreme Court held the patents invalid in that they were obvious in light of work by others. Justice Felix Frankfurter, dissenting, questioned the ability of his fellow justices to assess such "vast transforming forces of technology" from several decades removed.[17] It may not have helped the Marconi cause that Italy was by then at war with the United States.

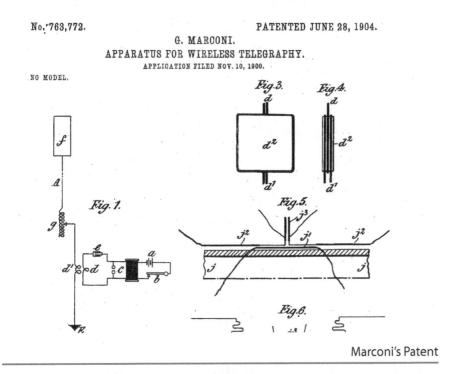

Marconi's Patent

The French and American governments battled over proper attribution for the discovery that HIV caused AIDS. French and American research groups had cooperated and shared material, and there was suspicion that the Americans had somehow misappropriated the virus. The governments eventually compromised. The leaders of the French and American teams were named as joint inventors on the patent on the test for AIDS. It has become generally accepted that there was no misconduct and that both teams contributed. The French isolated the virus, and the Americans proved its role in causing AIDS.

In 2008, however, the Nobel Prize in Physiology or Medicine went to two of the French teams in recognition of their work on the discovery of HIV, along with a cancer researcher. The resolution of the patent dispute evidently did not carry over to attribution for the scientific achievement.

Copyright law

Early American publishers found cheap inventory by reprinting English books without paying English authors. America shared the language (still does, even if *Harry Potter*'s American edition must change "nobbled" to "clobbered" and "queuing" to "lining up") and the books. By 1720, while still a colony, Massachusetts had five publishers reprinting English books,[18] soon to include works like Samuel Johnson's *Dictionary of the English Language* (1755).

Americans were not the first English-ruled territory to thrive on unauthorized copying of English books. Anglophile Irish printers in the 18th century had likewise feasted on books like *The Vicar of Wakefield* and *Tom Jones*.[19] This literary rebellion was followed by a literal rebellion in 1798. Unlike the American Revolution, this Irish rebellion was quelled, and many Irish publishers fled to America, where they pursued their previous trade in English wares. Copyright existed in America, but only for American authors.

The practice of unauthorized copying was widespread. Belgian publishers reprinted French works without permission, with only a portion of the population interested. However, starting with the Berne Convention of 1886, countries started to agree to honor the rights of each other's authors. The United States was slow to take interest, still gorging on works like those of Dickens.

Without U.S. Copyright, Dickens Gave Public Readings for Income (1862)

"Charles Dickens as he appears when reading." Wood engraving from a sketch by Charles A. Barry (1830–1892). Illustration in Harper's Weekly, v. 11, no. 571, 7 December 1867, p. 777, via Wikimedia Commons.

As the 20th century passed, though, America became more interested in the international protection of copyrights. American industries dependent on copyright sold their wares worldwide: movies, music, TV programs, software, video games. In addition, taking foreign works for free was good for U.S. publishers (cheap supply) and readers (cheap books), but it was doubly bad for U.S. authors, who had to compete with free books in the United States and did not get protection abroad. The United States gradually aligned its copyright law and became party to the Berne Convention, a "treaty for foreigners,"[20] under which countries promised to give foreign authors as good rights as domestic authors. If a country failed to live up to Berne, the consequences were—nothing. Berne had no enforcement provisions. "Stop, thief! I'll yell 'stop' again."[21]

By 1994 the United States joined TRIPS (Agreement on Trade-Related Aspects of Intellectual Property Rights), which was Berne with teeth: an international trade treaty under which countries, including the United States, promised to give certain rights in copyright, patent, trademark, and more. If a party fails to live up to TRIPS, international tribunals may award sanctions, such as limits on exports. Only a few countries are not party to TRIPS, Berne, or other copyright treaties with the United States: Afghanistan, Eritrea, Ethiopia, Iran, Iraq, San Marino, and Turkmenistan.[22] A U.S. citizen will have a British copyright in Britain, and a British subject will have an American copyright in the United States. The United States is one of the strongest proponents of international enforcement of copyright, patent, trademark, trade secrets, and the like, with considerable political heft.

The Congressional International Anti-Piracy Caucus issued its 2012 Piracy Watch List: China, Russia, Italy, Ukraine, and Switzerland. Switzerland a pirate haven? Switzerland is a landlocked country more noted for mountaineers than seafarers, aside from the Swiss Family Robinson. Evidently, the listing reflects different views on what the boundaries of copyright law should be. Switzerland, after conducting an extensive study that showed little effect from private downloading of sales of copyrighted work, has continued not to treat private copying as copyright infringement. The pejorative *pirate* can mean many things.

Pirate Flag?

Possible Early Pirate

Because the United States has agreed to international standards, U.S. copyright law is largely similar to that in other countries. But there are key differences.[23] Most countries do not have a copyright office. In the few that do, the copyright office is a pretty basic place, with only one function: to allow registration of a copyright. The U.S. Copyright Office, by contrast, has extensive systems for not only registering copyrights but recording copyright transactions, holding hearings, and issuing regulations on copyright issues.

Also, most countries do not have one broad rule of fair use but rather have more specific rules, such as allowing photocopying by teachers. The penalties for infringement in the United States tend to be greater.

Other countries are more likely to have *moral rights*, special rules to protect artists. The difference is reflected in the names of the counterparts of copyright in Europe, such as *droit d'auteur* (rights of author, France) and *Urheberrecht* (creator's right, Germany). Such rules may require payment to the author if a painting is resold or a book becomes a smash hit (Der "Bestseller-Paragraph"[24]). Artists may have rights to get credit and to prevent distortion of their work.

Questions

1. Portia writes her first novel. What does she have to do to have copyright across the globe?
2. Turner Network Television did great business by colorizing black-and-white movies for a new generation of viewers. But not in France. Why not?
3. In Samuel Beckett's *Waiting for Godot*, published in Paris in 1952, nothing really happens. Al writes a parody that is within the bounds of fair use. Any risks?

Answers

1. She has done it already. The copyright laws of almost all countries recognize the copyrights of foreign authors. Nor does copyright depend on formalities like registration, use of a copyright notice, and deposit of copies.
2. French courts, more protective of author's rights than U.S. courts, held that colorization was not an adaptation permitted by ownership of the copyright but instead a distortion of the work that would offend the moral rights of the original author.
3. Al might not infringe under U.S. copyright law. But counties that weigh heavier toward protecting works of artists might call it infringement. Al should get advice before he publishes or performs the parody in France, for instance.

Patent law

Early patents were, in part, a tool of mercantilism, which was the policy to exclude foreign businesses by means of high tariffs on imports, control of trading by colonies, and other tactics. Patents for domestic businesses (which were often granted to government cronies rather than inventors) could thus exclude foreign products. Nations learned that mercantilism was mutually damaging to their economies. In the move toward more free trade, patent law likewise changed. The United States was quicker in patent law than in copyright law to extend protection to foreigners; patent protection was provided as early as 1800. The major European countries and the United States all agreed to protect foreign inventions with the Paris Convention of 1883. Patent is now also part of TRIPS, so international standards govern. Like the United States, other countries struggled with questions such as who should decide what is patentable. In the United Kingdom the Royal Society, a group of eminent scientists, sought that role for a time.

But in the field of patent law, the United States was more ready to compete. Americans were early contributors to technological innovation, and patent law reflected that. Domestically, the United States was quick to implement a thorough patent law and an effective office to administer the law. Internationally, the United States quickly joined international agreements and gave patent protection to foreign inventors. The patent records reflect the widespread impulse in the United States to invent—and to seek commercial advantage from patenting inventions. The number of patents per capita was greater in the United States than elsewhere. Influential figures often credited the U.S. patent system for the widespread technological advances in the nation, along with its economic benefits.

U.S. patent law did differ from the systems in Europe.[25] As Zorina Khan's research has shown,[26] the differences both reflected and reinforced the democratic nature of the United States, in several senses of the word. In Europe, patents were not given so freely but rather were reserved for special inventions. Moreover, patent fees were high, making it impracticable for the average inventor to secure legal rights. The national patent offices were strongly influenced by aristocratic privilege, meaning that social connections were often more important in securing rights than individual inventive contributions. One French applicant located by Khan included the evidently relevant information that his wife was a wealthy heiress and gave her five first names and the name of her noble family on his patent application. His application was granted. Drawing on records from the respective patent systems, Khan showed that patentees in Europe were more likely to be from the elite classes than those in the United States. They were also more likely to be from major cities as opposed to poorer rural regions.

Patents operated quite differently in the young United States than in these other countries, Khan showed. The patent office was in some senses one of the most democratic institutions in the country. Rather than a place of patronage, the patent office was subject to typically American checks and balances. This resulted in greater confidence in the office, with a marked difference in controls on patents in comparison with other countries. In some European countries, there was no system of patent examination for fear that examiners would extract favors in exchange for favorable rulings. Indeed, patents began simply as prerogatives granted by royalty. But because the USPTO was relatively politically independent and trustworthy, patent applications were examined before patents were issued.

Other limits were also absent in the United States. Patent fees were kept far lower than in other countries, allowing far greater social access to patent rights. Even the cost of mailing the application was spared the inventor because the U.S. Post Office gave free postage to patent applications. The standard for patent protection was also lower. Rather than reserving patents for exceptional inventions, patents were granted for even modest contributions to homely technologies.

The USPTO was also *relatively* more open than other governmental institutions. Khan's research shows that inventors were not barred by race or gender from applying for patents. Free blacks secured patents as early as 1821, although slaves were still denied the ability to patent their inventions. Women regularly received patents, although family and property law often denied them the ability to fully exploit their inventions commercially. Low patent fees and a straightforward examination process meant that lack of wealth, education, or connections would not bar patent protection.

Considerable differences remain between the United States and other countries. Patentable subject matter varies quite a bit. Software patents are more difficult to get in Europe than in the United States, although clever patent drafters find ways to characterize software devices as machines. India did not permit patents on pharmaceuticals until recent years. Such patents are subject to stronger policing there than elsewhere. The Indian patent office denied Novartis a patent on its blockbuster cancer drug, Glivec®. The application was held to be "evergreening," which means seeking a patent on a minor change to an existing drug.[27] In the United States, by contrast, Glivec was deemed to be patentable, not an obvious improvement. India also applies compulsory licensing more often than in the United States. Bayer sold Nexavar® for kidney cancer there but did not sell much because the price of $5,000 per month was many times the per capita income. An Indian firm was granted a compulsory license (meaning Bayer's agreement was not required) by the government to sell the drug at some 3 percent of the price, with a percentage royalty to Bayer. (It should be noted that some international pharmaceutical companies have made considerable efforts to make important medicines available in developing countries.)

The friendly sounding Patent Cooperation Treaty makes it simpler to seek patent protection in multiple countries. An inventor can file the patent application in his home country and simply mark it to be forwarded to specified countries.

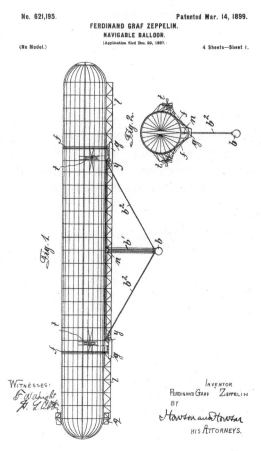

Zeppelin's U.S. Patent

For Microsoft to patent in Micronesia, it need not go to Pohnpei Island, once Micronesia joins the PCT. Microsoft can simply mark "Micronesia" when it files its U.S. patent application and pay the extra fee. There is no international patent. Each country applies its own standards. Micronesia would decide if Microsoft's holodeck is patentable there. Meanwhile, Microsoft can make a virtual visit to Pohnpei Island using the invention.

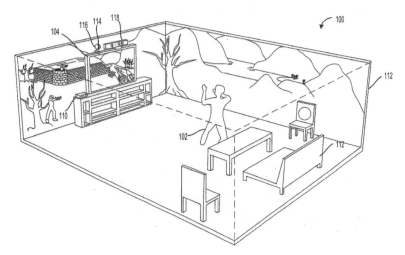

From Microsoft U.S. Patent Application on "Immersive Display Experience"[28]

Seeking patents in numerous countries is expensive. Each country charges fees. Many national patent offices require the application to be translated into their language, which adds considerably to the cost of filing international applications. Not all countries require this, though. Among European countries, Iceland requires only the claims and abstract to be translated into Icelandic and allows the description in English. This is a little ironic, because Icelandic borrows English words less than other languages do, instead creating words with indigenous roots. Google Translate renders "genetics, algorithm, and computer" as "erfðafræði, reiknirit, and tölva." Meanwhile, the extensive genetic and genealogical computer databases of the Icelandic population have produced U.S. patents on genes—and on algorithms for searching big databases.

(19) **United States**

(12) **Patent Application Publication**   (10) Pub. No.: **US 2010/0174714 A1**
    Asmundsson et al.   (43) **Pub. Date:**   **Jul. 8, 2010**

(54) **DATA MINING USING AN INDEX TREE CREATED BY RECURSIVE PROJECTION OF DATA POINTS ON RANDOM LINES**

(75) Inventors: **Fridrik Heidar Asmundsson**, Mosfellsbaer (IS); **Herwig Lejsek**, Scheibbs (AT); **Bjorn Thor Jonsson**, Reykjavik (IS)

Correspondence Address:
Muncy, Geissler, Olds & Lowe, PLLC
4000 Legato Road, Suite 310
FAIRFAX, VA 22033 (US)

(73) Assignee: **HASKOLINN I REYKJAVIK**, Reykjavik (IS)

(21) Appl. No.: **12/303,598**

(22) PCT Filed: **Jun. 6, 2007**

(86) PCT No.: **PCT/IS07/00014**

§ 371 (c)(1),
(2), (4) Date: **Apr. 3, 2009**

(30) **Foreign Application Priority Data**

Jun. 6, 2006 (IS) 8499

Publication Classification

(51) **Int. Cl.**
G06F 17/30 (2006.01)

(52) **U.S. Cl.** **707/737**; 707/E17.089; 707/776; 707/752; 707/797; 707/E17.012

(57) **ABSTRACT**

The present invention relates to a method computer program product for datamining with constant search time, the method and computer program product comprises the steps of: traversing a search tree to a leave, retrieving a one or more data store identifier from said leave, read data pointed to by said data store identifier, locating one or more value in said data, referencing one or more data descriptor, retrieve the n-nearest data descriptor neighbors, terminate said search.

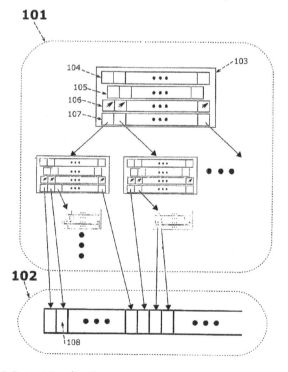

101

102

108

Icelanders' U.S. Patent Application

The cost of international patents deters many inventors. Edison, who saw little use for moving pictures, got a U.S. patent for his movie projector but did not spend the money for patents in other countries—to his regret.

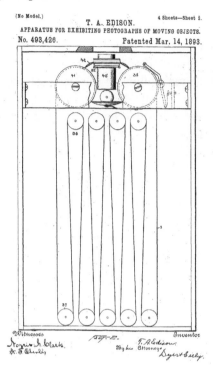

Edison's U.S. Movie Camera Patent

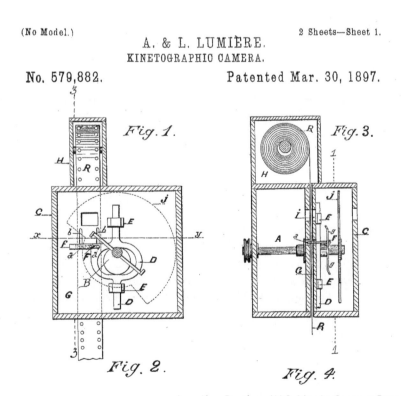

Lumière Brothers' U.S. Movie Camera Patent

Trademark and patent law

Trademarks affect national images. Consider Samsung and Hyundai for South Korea; Hermès, Moët & Chandon, and Cartier for France; Ferrari, Fiat, Armani, and Gucci for Italy; Tata and Amul for India; Coca-Cola, Google, and McDonald's for the United States. Regional images likewise: in Cumbria, home of the beauteous Lake District, some take much pride in the Honister Slate Mine and the innovations of the Derwent pencil company.

Trademarks get broad mutual recognition of trademark rights under TRIPS. In addition, the Madrid Protocol created a mechanism to launch international applications from a single domestic application. As brands are increasingly sold globally, international protection is key to entering markets and preventing knockoffs that might make their way back to the mark owner's home country.

International harmonization still has some sour notes. There are business and legal pitfalls. Understanding how symbols will be interpreted in another country is key to using a mark there. Some missteps collected by a commentator are as follows: the famous American Dairy Association slogan "Got Milk?" was translated for the Mexican market as "Are You Lactating?" "Software" has been translated as "underwear." Finns have considerable experience in dealing with frozen locks on car doors in winter. They exported the results of their consequent expertise to the United States in a product called Super Piss. The product, however, did not get shelf space in U.S. stores, perhaps reflecting cultural differences among storekeepers.

Geographic appellations

> *Mousebender*: Cheshire?
> *Wensleydale*: No.
> *Mousebender*: Any Dorset Blue Vinney?
> *Wensleydale*: No.
> *Mousebender*: Brie, Roquefort, Pont-l'Évêque, Port Salut,
> Savoyard, Saint-Paulin, Carre-de-L'Est, Boursin, Bresse-Bleu,
> Perle de Champagne, Camembert?
> *Wensleydale*: Ah! We do have some Camembert, sir.
> *Mousebender*: You do! Excellent.
> *Wensleydale*: It's a bit runny, sir.
>
> Monty Python, "The Cheese Shop"

Countries with a history of local agriculture and artisan foodstuffs have long given more protection to marks tied to a particular location. The United States traditionally protected geographic marks only if the particular producer had achieved renown. Marks that referred to a product made by many producers from a particular region (think of Champagne) were often treated as generic marks. The United States is gradually moving to harmonize with the protection of geographic appellations.

Gray market (a.k.a. parallel importation)[29]

Sellers often sell at different prices in different markets. A physics textbook might sell for a high price to U.S. college students while a low-cost edition with the same content may be published and sold at a lower price abroad. An entrepreneur might buy low abroad and sell high in the United States, taking advantage of the price difference to get a profit from arbitrage. Some sellers use eBay; others use container ships. The trademark owner will argue that this is likely to confuse consumers, who will associate the mark with the higher quality U.S. version of the product. If the product is genuine, however, and is in mint condition, it likely does not infringe trademark.

Question

Suppose a book or drug sells for a high price in one country and a low price elsewhere. Should the copyright or patent owner be able to maintain separate markets, or should it be legal to buy low, export, and sell high?

Answer

The copyright or patent holder will argue that price discrimination allows it to reach more buyers, which is good for everybody. If arbitrage is allowed, it may simply stop selling in the low-price country. Others will argue that once sold, the book or drug should no longer be controlled by the copyright or patent. What do you think?

Indigenous Peoples and Intellectual Property[1]

Yankee Crafters of Cape Cod registered the trademark WAMPUM JEWELRY®. Reportedly, Yankee's owner vigorously asserted rights to the mark, even against the Mashpee Wampanoag tribe, makers of wampum jewelry centuries before the arrival in Cape Code of Yankees, Puritans, Pilgrims, or tourists: "If the tribe wants the word back, they can buy it from me."[2]

Urban Outfitters'"Ecote Navajo Wool Tote Bag"[3]

The Wapishana in Guyana told a chemist of their medical uses of the greenheart tree. He isolated the active ingredient, named it *rupununine* after a local river, obtained a U.S. patent, and sought investors. As he noted, allocating rights is not clear-cut. "Tough, isn't it? . . . How can I tell the Wapishana about the science? They just inherited the greenheart. They don't own it. I have invested in this with my own money."[4]

United States Patent [19]

Gorinsky

[11] Patent Number: **6,048,867**

[45] Date of Patent: **Apr. 11, 2000**

[54] **BIOLOGICALLY ACTIVE RUPUNUNINES**

[76] Inventor: **Conrad Gorinsky**, The Old House, Old House Lane, Nazeing, Essex ENG225, United Kingdom

[*] Notice: This patent issued on a continued prosecution application filed under 37 CFR 1.53(d), and is subject to the twenty year patent term provisions of 35 U.S.C. 154(a)(2).

[21] Appl. No.: **08/689,824**

[22] Filed: **Aug. 12, 1996**

Related U.S. Application Data

[63] Continuation-in-part of application No. 08/189,781, Feb. 1, 1994, Pat. No. 5,569,456.

[51] Int. Cl.[7] **A61K 31/4741**; A61P 35/00
[52] U.S. Cl. **514/279**; 514/25; 514/30; 514/308; 514/841; 514/895; 514/918; 514/919; 546/35; 546/140
[58] Field of Search 514/279, 308, 514/841, 895, 918, 919, 25, 30; 546/35

[56] **References Cited**

PUBLICATIONS

Marshall SJ et al. Antimicrobial Agents and Chemotherapy, 38(1) pp. 96–103, Jan. 1994.
Cortes D et al. Journal of Natural Products, 55(9) pp. 1281–1286, Sep. 1992.

Ivorra MD et al. European Journal of Pharmacology, 219 (2), pp. 303–309, 1992.

Primary Examiner—Evelyn Mei Huang
Attorney, Agent, or Firm—Nixon & Vanderhye

[57] **ABSTRACT**

A method of therapy wherein there is administered to a person in need of such therapy a pharmacologically effective amount of a rupununine having the formula:

wherein R = ——H or ——CH₃.

3 Claims, No Drawings

Greenhart Patent

Not Really from the Kickapoo Tribe

Amazonian people used the powerful hallucinogenic plant ayahuasca for ceremonial and medical purposes. An American scientist learned of the plant's qualities from them and obtained a plant patent on a new variety. The patent was invalidated based on information from shamans and museums.[5]

United States Patent [19]

Miller

[11] Patent Number: Plant 5,751
[45] Date of Patent: Jun. 17, 1986

[54] *BANISTERIOPSIS CAAPI* (cv) 'DA VINE'

[76] Inventor: Loren S. Miller, 1788 Oak Creek Dr., Apt. 407, Palo Alto, Calif. 94303

[21] Appl. No.: 669,745

[22] Filed: Nov. 7, 1984

Related U.S. Application Data

[63] Continuation of Ser. No. 266,114, May 21, 1981, abandoned.

[51] Int. Cl.⁴ A01H 5/00
[52] U.S. Cl. Plt./54
[58] Field of Search Plt./54

[56] References Cited

U.S. PATENT DOCUMENTS

P.P. 3,008 12/1970 Magnuson Plt./88
P.P. 4,253 5/1978 Arnold Plt./88

OTHER PUBLICATIONS

Gates, Bronwen *Flora Neotropica* Monograph No. 30, Banisteriopsis, Diplopterys (Malpighiaceae) Published for Organization for Neotropica by the New York Botanical Garden, N.Y. Feb. 18, 1982.

Primary Examiner—James R. Feyrer

[57] ABSTRACT

A new and distinct *Banisteriopsis caapi* plant named 'Da Vine' which is particularly characterized by the rose color of its flower petals which fade with age to near white, and its medicinal properties.

2 Drawing Figures

Ayahuasca Patent

A U.S. patent was obtained on a slight variation of a bean used for centuries by traditional farmers in Mexico. The patent was challenged and invalidated, but only after a decade, not long before it would expire anyway, and at considerable cost to the bean's market.[6]

U.S. Patent   Apr. 13, 1999   Sheet 1 of 3   5,894,079

Figure 1

Enola Bean Patent

A medical research team took blood samples from the Ngäbe of Panama on the pretext of battling a fictitious infectious disease. They discovered that one woman had natural resistance to leukemia and sought a patent on material from her cell line.[7] A U.S. patent was similarly obtained for a T-cell line from a blood sample of the Hagagai over the objections of the government of Papua New Guinea.[8]

US005397696A

United States Patent [19]

Yanagihara et al.

[11] Patent Number: 5,397,696

[45] Date of Patent: Mar. 14, 1995

[54] **PAPUA NEW GUINEA HUMAN T-LYMPHOTROPIC VIRUS**

[75] Inventors: **Richard Yanagihara; Vivek R. Nerurkar**, both of Frederick, Md.; **Carol Jenkins**, Goroka, Papua New Guinea; **Mark Miller**, Fort Lee, N.J.; **Ralph M. Garruto**, Boyds, Md.

[73] Assignee: **The United States of America as represented by the Department of Health and Human Services**, Washington, D.C.

[21] Appl. No.: **743,518**

[22] Filed: **Aug. 12, 1991**

Related U.S. Application Data

[63] Continuation-in-part of Ser. No. 572,090, Aug. 24, 1990, abandoned.

[51] Int. Cl.6 .... **C12N 7/02**
[52] U.S. Cl. **435/5; 435/235.1; 435/7.1; 435/7.2; 435/7.21; 435/7.24; 435/7.92; 435/239**
[58] Field of Search 435/5, 7.1, 7.2, 7.21, 435/7.24, 7.92, 235.1, 239, 237

[56] **References Cited**

PUBLICATIONS

Asher et al: "Ab to HTLV–I in Populations of the Southwestern Pacific" J. of Med Vir 26:339–51 (1988).
Popovic et al "Transformation of Human Umbilical Cord Blood T–cells by HTLV" PNAS 80:5402–6 (1983).
Gallo et al "Comparison of Immunofluorescence, EI, & WB Methods for Detection of Ab to HTLV–1" J. of Clin Microb V26 #8 pp. 1487–1491 (1988).

Primary Examiner—Christine M. Nucker
Assistant Examiner—Jeffrey Stucker
Attorney, Agent, or Firm—Morgan & Finnegan

[57] **ABSTRACT**

The present invention relates to a human T-cell line (PNG-1) persistently infected with a Papua New Guinea (PNG) HTLV-I variant and to the infecting virus (PNG-1 variant). Cells of the present invention express viral antigens, type C particles and have a low level of reverse transcriptase activity. The establishment of this cell line, the first of its kind from an individual from Papua New Guinea, makes possible the screening of Melanesian populations using a local virus strain. The present invention also relates to vaccines for use in humans against infection with and diseases caused by HTLV-I and related viruses. The invention further relates to a variety of bioassays and kits for the detection and diagnosis of infection with and diseases caused by HTLV-I and related viruses.

5 Claims, 33 Drawing Sheets

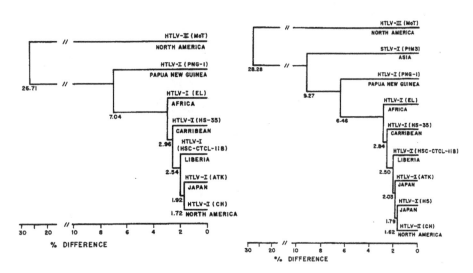

Papua New Guinea Human T-Lymphotropic Virus

A U.S. patent was obtained on putting turmeric on a wound, an Ayurvedic remedy for centuries in India. The patent was later invalidated after submissions surfaced including "ancient Sanskrit text and a paper published in 1953 in the *Journal of the Indian Medical Association*."[9] The Indian government, in reaction to the many Western patents, copyrights, and trademarks on yoga practices and materials, has begun compiling information, including translations of sacred documents, to show that such information was already in the public domain.[10]

United States Patent [19]

Das et al.

[11] **Patent Number:** **5,401,504**

[45] **Date of Patent:** **Mar. 28, 1995**

[54] **USE OF TURMERIC IN WOUND HEALING**

[75] Inventors: **Suman K. Das; Hari Har P. Cohly,** both of Jackson, Miss.

[73] Assignee: **University of Mississippi Medical Center,** Jackson, Miss.

[21] Appl. No.: **174,363**

[22] Filed: **Dec. 28, 1993**

[51] Int. Cl.⁶ **A61K 35/78**

[52] U.S. Cl. **424/195.1;** 514/925; 514/926; 514/927; 514/928

[58] Field of Search 424/195.1

[56] **References Cited**

U.S. PATENT DOCUMENTS

4,719,111 1/1988 Wilson 424/195.1
5,120,538 6/1992 Oei 424/195.1
5,252,344 10/1993 Shi 424/682

OTHER PUBLICATIONS

Institute GA. 99: 218620T (1983) of JPN. 58–162520

(Ulcer Inhibitor Tablets Effective in Mice Contain Carcinogen).
Soma et al GA. 116: 221612S (1992) of JPN 4–49240 (Digestive Tract Ulcers Treated with *Curcuma longa* (Turmeric) Extract (Lipopolysaccharides)).
Kumar et al GA.119: 871K (1993) of Ind. Vet. J. 70(1):42–4 (1993).
Abstracts of Charles et al Trop. Geogr. Med: 44(1–2) 178–181 Jan. 1992; Rafatullah et al J. Ethnopharmacol. 29(1): 25–34 Apr. 1990; Kutton et al Tumori 73(1): 29–31 Feb. 28, 1987; Mehra et al. Tokai J Etpharm Med 9(1): 27–31 Mar. 1984.

Primary Examiner—Shep K. Rose
Attorney, Agent, or Firm—Wenderoth, Lind & Ponack

[57] **ABSTRACT**

Method of promoting healing of a wound by administering turmeric to a patient afflicted with the wound.

6 Claims, No Drawings

Turmeric Patent

Pablo Picasso drew heavily on African art. Picasso's *Les Demoiselles d'Avignon*, considered the first cubist work, "drew on sources as diverse as Iberian sculpture, African tribal masks, and El Greco's painting to make this startling composition."[12]

"Reliquary Head (Nlo Bieri), 19th–20th Century, Gabon; Fang, Betsi Group," Metropolitan Museum of Art, New York[11]

Picasso's Universal Remix

Do you think copyright, patent, trademark, or trade-secret law should prevent any of the aforementioned, or should the information be free?

The following gives some things to think about on that broad question. No doubt you can think of more.

No IP for IP?

Copyright, patent, trademark and trade-secret laws offer little protection for traditional knowledge, folklore, traditional songs and dances, and other elements vital to the culture of indigenous people. Copyright, for example, protects only the work of identified authors for limited periods of time. Songs or dances that are part of the cultural heritage of a group would run afoul of those limits. Traditional knowledge of medicines and other technology may likewise be important to an indigenous group. But patent law will not protect knowledge when the original inventor is long gone and the product or process has been in use.

As for trademarks, indigenous groups have not registered their symbols as trademarks (registration is necessary in many countries besides the United States). Sacred or other cultural uses may often not qualify as a trademarkable commercial use. And as to trade secrets, some knowledge may be freely shared. Even guarded knowledge may not be subject to the sort of security measures necessary for trade-secret protection. The information may also not fall within the protected category of information with commercial value. Certainly indigenous groups often have not had the political or legal means to use even the limited assistance the law might give.

Conversely, as the preceding examples show, outside parties may use copyright, patent, trademark, or trade-secret law, sometimes in a questionable way, to exploit the information of indigenous groups. Copyright will apply if just a smidgen of creativity is added to an existing work. Patents may be had on an improvement of an existing technology or isolation of a natural phenomenon. Symbols may be appropriated as trademarks. Traditional knowledge known among an indigenous group can become a trade secret if taken elsewhere and used in an industry previously unaware of the information.

Questions

1. An indigenous group uses its traditional knowledge that a certain mixture of herbs effectively treats fatigue. Why might the group not be able to patent that information? Biochemists learn of the herbs and isolate the active ingredients. Why might the biochemists be able to patent that information?
2. The Ami have passed on their "Song of Joy" for many generations. Can they copyright it? German rock group Enigma uses a recording of the "Song of Joy" as part of "a popular world-beat tune known as 'Return of Innocence.'"[13] Can Enigma copyright it?

1. Patenting may require legal and financial resources that the group does not have. Beyond that, the original inventor may have died generations ago. If the information has been in use, it may no longer be novel and so is not patentable.

   Someone who learns the information from the group cannot patent it because he did not invent it. But if he isolates it or develops a drug based on it, that may be sufficiently new to patent.

2. The original author (and authors of adaptations) may no longer be known. The song may be too old to be copyrighted.

   An adaptation of a work may be copyrighted, so the electronic dance song may qualify for its own copyright, without any rights going to the Ami. Should the Ami have any legal rights here?

International pacts: IP for IP for IP

International law presently provides the most recognition of the imbalance between the strong protection of commercial "intellectual property" and the lack of protection for indigenous groups.[14] A United Nations report on the "Protection of the Heritage of Indigenous People" emphasizes the link between protection of indigenous heritages and the right of indigenous peoples to exist as "distinct peoples" in control of their own destinies:

> The protection of cultural and intellectual property is connected fundamentally with the realization of the territorial rights and self-determination of indigenous peoples. Traditional knowledge of values, autonomy, or self-government, social organization, managing ecosystems, maintaining harmony among peoples and respecting the land is embedded in the arts, songs, poetry and literature which must be learned and renewed by each succeeding generation of indigenous children.[15]

The United Nations Declaration on the Rights of Indigenous Peoples states,

> Indigenous peoples have the right to maintain, control, protect and develop their cultural heritage, traditional knowledge and traditional cultural expressions, as well as the manifestations of their sciences, technologies and cultures, including human and genetic resources, seeds, medicines, knowledge of the properties of fauna and flora, oral traditions, literatures, designs, sports and traditional games and visual and performing arts. They also have the right to maintain, control, protect and develop their intellectual property over such cultural heritage, traditional knowledge, and traditional cultural expressions.[16]

The Declaration by itself does not give the right to enforce such rights in court. Rather, individual countries may implement it by changes in their own laws.

Cut into public domain?

Some have raised the concern that giving legal protection to folklore or traditional knowledge would remove matter from the public domain.[17] But many indigenous claims are to prevent others from privatizing indigenous knowledge.[18] Folklore or indigenous music, for example, are often not protected by copyright. But a modern recording containing indigenous music or a new book containing folklore may be copyrighted if it contains even minimal new elements (the choices made in the sound recording or the editorial additions to the folklore). So a commercial interest may, in effect, be able to hold copyrights in the products of an indigenous group. Similarly, in biotechnology, traditional knowledge about the properties of plants can legally be privatized by outsiders. Patent law "enables broad patents on minor modifications, syntheses, and purifications,"[19] such as "plant genetic resources, in which patents based on local knowledge of plant qualities have become routine."[20] Traditional knowledge may not qualify as a trade secret if not subject to security measures or if not used for commercial value.[21]

The public domain may benefit some more than others. In theory, the public domain is a rich resource of ideas available to all. A vibrant public domain is essential to creativity in every field. But expansion of the public domain may mean different things, especially in the case of indigenous peoples.[22] In particular, "differing circumstances—including knowledge, wealth, power, and ability—render some better able than others to exploit a commons."[23] If such elements as traditional knowledge and folklore are completely within the public domain, the commercial interests able to exploit them most efficiently could benefit most. The present balance of intellectual property law gives protection to the knowledge generated by developed countries while tending to leave open to everyone the knowledge generated by developing countries and indigenous peoples. Adjusting that balance would likely be a move toward fairness. As the UN report concludes, international and domestic measures may well be necessary in order for "indigenous peoples to retain control over their remaining cultural and intellectual, as well as natural wealth, so that they have the possibility of survival and self-development."[24]

Group rights[25]

Much has been made of the apparently individualistic bent of intellectual property law. Group rights, by contrast, have been deemed alien to the basic framework of such laws. But there are many areas in which intellectual property rights are both created and held by groups. A movie may be the product of creative contributions from dozens or even hundreds of directors, writers, actors, costumers, special effects technicians, and more. Inventions from pharmaceuticals to software to aeronautics all require input from many people over a long period of time.

Trademarks often act as a banner giving an illusion of unity to vast international enterprises. The rights to copyrights and patents are often held by corporations with millions of shareholders. Trademark law also serves groups. A trademark can be a unifying symbol for a vast enterprise—think only of Coca-Cola, McDonald's, or the U.S. Postal Service.

Furthermore, particular types of trademarks can specifically serve to promote the interests of groups. A union can use a collective mark. A certification mark can be used to demonstrate that a product or service conforms to the standards shared by a group. The "OSI Certified" mark can be used on software to show that it conforms to the principles of the open source software movement—a large, dispersed group united mainly by certain beliefs about computer programs. Accordingly, trademark law can also fit into protecting group rights with respect to indigenous peoples.

Intellectual property in this case would not serve its usual role of providing economic incentives. Rather, cultural property can provide key elements to support other rights, such as the rights to education, to practice cultural traditions, and to self-determination.[26] If a filmmaker recorded the performance of a sacred dance or a folk talk, the filmmaker would have copyright in the work while the indigenous group would have no say about dissemination of the work because their contribution would be deemed not creative. Evening out some of those disparities would support the rights of self-determination for indigenous people—and in a broader sense, it would support the justifications to have intellectual property. The public domain can be misleading if the effect is to open the cultural heritage of one group for the use of other groups.[27]

Respect for cultural identity and self-determination need not entail the strong property rights often associated with "intellectual property."[28] Indigenous people could be consulted before others made use of their information or symbols, participate in the commercialization of their traditional knowledge, and prevent deceptive uses of cultural property.

Some indigenous people have taken the initiative. The Quileute Nation saw its name and traditions mixed, without consultation, with vampire lore in the blockbuster *Twilight* books and movies. The Quileute adapted their existing tourism businesses accordingly:

> Come experience the tranquility and natural beauty of the rugged coastline of the mighty Pacific and visit the haunts of all your favorite characters from the popular book series penned by Stephenie Meyer. Jacob Black and the rest of the Quileute Tribe invites you to experience La Push and 1st Beach.[29]

Such efforts support economic development and can also guide discussion toward more accurate portrayal of cultural elements.

Forks, Washington

Good Fences Make Good Neighbors (Actually, Just a Jest)

Cooperation may be fostered by respect for culture and traditional knowledge. The Blackfeet of Montana operate Sun Tours in Glacier National Park, showing "Glacier's natural features relevant to the Blackfeet Nation, . . . commonly used plants and roots for nutrition and medicine are pointed out."[30] The National Park Service supports the tours and offers related educational curricula and cultural events.

In India,

> the medicinal knowledge of the Kani tribes led to the development of a sports drug named Jeevani, an anti-stress and anti-fatigue agent, based on the herbal medicinal plant arogyapacha. Indian scientists . . . isolated 12 active compounds from arogyapacha, developed the drug Jeevani, and filed two patent applications on the drug. A trust fund was established to share the benefits arising from the commercialization of the TK-based drug.[31,32]

Cooperation, consultation, and respect benefit all.

Internet IP: Brave New World[1]

The law must adjust to changes in society and its technology. It is often said that copyright law was written for books and music and did not contemplate today's networked world. The Internet is a worldwide machine for making and distributing copies. But we can exaggerate how unforeseeable our times were.

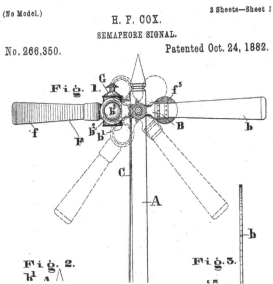

France Had a Semaphore Network in the 1800s

The United States Had Pneumatic Tube Mail Networks in 1900 in New York (27 Miles!) and St. Louis[2]

Benjamin Kaplan's 1967 book *An Unhurried View of Copyright* analyzed how copyright law would have to adapt to a future worldwide computer network with multimedia capabilities. His vision of a Web beats many conjectured cyberspaces:

> You must imagine, at the eventual heart of things to come, linked or integrated systems or networks of computers capable of storing faithful simulacra of the entire treasure of the accumulated knowledge and artistic production of past ages, and of taking into the store new intelligence of all sorts as produced.[3]

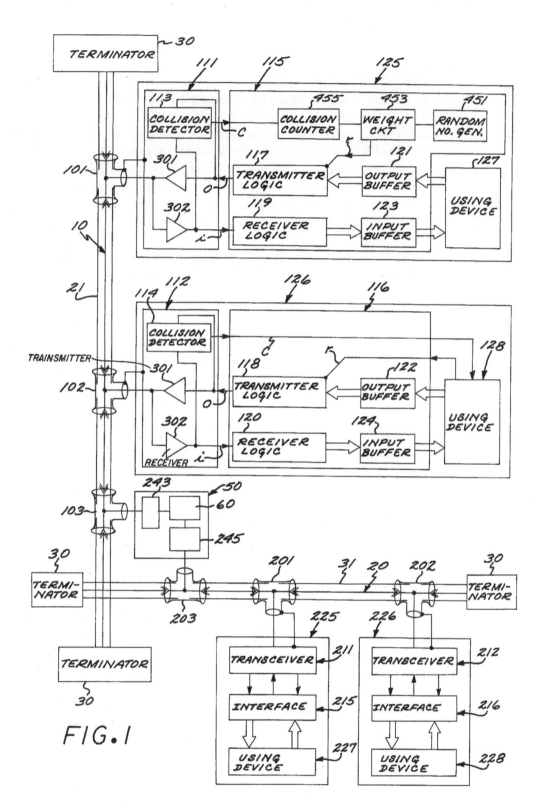

FIG. 1

Robert M. Metcalfe's Patent on Ethernet: Much Early Internet Technology Was Not Patented, Such as Internet Protocol (IP) and Transmission Control Protocol (TCP), in Part because the Research Was Government Funded

Kaplan foresaw many of today's issues. The ease with which one can copy, adapt, and distribute works raises a tension with the "'moral rights' of authors to prevent abuses in the exploitation of their creations."[4] Kaplan spoke of the need to encourage a registration system in which authors could submit their works in a form permitting indexing, abstracting, storage, and retrieval—an area where the copyright office struggles to keep up. If that could be done, Kaplan observed, it would no longer be necessary to publish academic material in "a mélange of learned journals and in the output of university presses."[5] Such material would be available online—both manuscripts that had been accepted for "publication" and those that had not. For more commercial works, there would likely be two phases: an initial broad commercialization phase followed by a secondary phase with what would now be called digital rights management systems: "book-keeping apparatus that can continue for the whole copyright period to bill the customers monthly or weekly" while "preventing unconsented-to private copying of works."[6]

There would be issues about how such systems would be administered, Kaplan foresaw, and about whether the government would have a role to play. All that foreshadowed such controversies as the Google Book project, legal protections for anti-copying technology, online access to academic articles (e.g., JSTOR), and the ongoing negotiations for amendments to copyright treaties to deal with issues of media "piracy." Kaplan also saw issues with respect to "automatic translation, in world-wide networks."[7]

FIG. 2

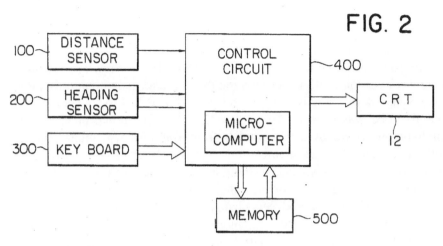

FIG. 3

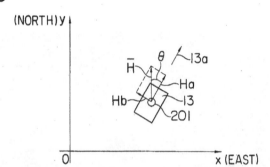

FIG. 4

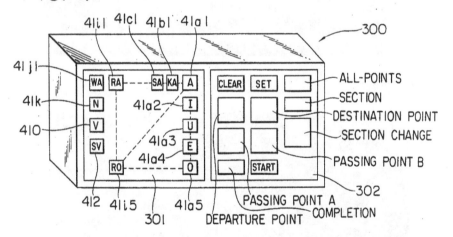

The First U.S. Patent Using the Word *Internet* on an
"Automotive Navigation System"

Immunity for Internet service providers

In the German film *Der Untergang* (*Downfall*), Hitler reacts with rage when he learns that troops of the USSR are closing in on Berlin. The film, subtitled in English, had modest U.S. box office and DVD sales. A YouTube user posted a clip of the film after editing in his own subtitles so that Hitler was reacting to the lame features in Microsoft's Flight Simulator X. Soon there were versions with Hitler learning that the iPad cannot multitask, that Santa Claus does not exist, that Manchester United lost to Barcelona, and that YouTube was showing Hitler parodies. At some point, however, the film's producers removed the parodies from YouTube. Up sprang clips of Hitler reacting with rage upon learning that the Hitler parody videos were being taken down. Some made a strong argument that fair use protected the parodies. *Untergang*'s director even spoke admiringly of the wit in some of the parodies. The film's producers changed tactics, permitting the videos to stay up—along with ads, whose revenue went in part to the film's producers. None of this required going to court. How so?

Der Untergang, Remixed

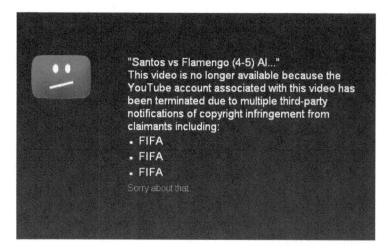

Taken Down and Terminated

As has been said before, the Internet is a worldwide copying machine. Many Internet service providers (ISPs), like YouTube and Facebook and Wikipedia, host content provided by users. If the content infringes, the ISP potentially infringes. In the United States, copyright law provides an ISP with immunity for copyright infringement by its users. So YouTube does not infringe for material posted by its users and for copies made in network operations. Immunity has its price, though. An ISP must take down material that it knows is infringing and terminate repeat infringers. The ISP must also have a program to receive and comply with take-down notices from copyright owners.

The law has some safeguards for users. When material is taken down, a user may send a counter take-down notice. The ISP must then put the material back up unless the copyright owner files an infringement suit within ten days. A copyright claimant can itself pay damages if it sends a take-down notice with no good reason. Stephanie Lenz posted a YouTube video of her toddler dancing to the tune of Prince's "Let's Go Crazy" playing in the background. The music company soon sent a take-down notice. But Lenz's use was fair use, as an incidental, noncommercial use with no market harm. The music company could have to pay for sending the take-down notice with no basis. Other notorious baseless take-downs were brought by the maker of voting machines trying to squash analysis of the machines' software, and a psychic trying to prevent analysis of his telekinetic spoon-bending.[8]

Much of take-down practice is now automated, as is much copyright infringement. The software of copyright holders searches the Internet and generates take-down notices when it detects the content of its masters. The ISP's computer receives the take-down notice and removes the material, all without human intervention.

The Hitler parodies, however, were not taken down as a result of take-down notices, nor were they put back up as a result of counter take-down notices. Copyright law's take-down procedure, with its safeguards for users, has been somewhat displaced by private arrangements. YouTube permits copyright owners to upload files to its Content ID system. Videos posted by users are compared to the Content ID files. If there is a match, the copyright owner may block the video, see its viewing statistics, or allow advertisements to be placed with the video. Advertising revenue from such videos is a big portion of YouTube's revenue. With the Hitler parodies, Constantin Film initially chose to block the videos, then later chose to leave them up, but with advertisements. Music companies likewise often let their content be used in exchange for advertising revenue.

NASA posted a video on YouTube about its rover Curiosity landing on Mars. Scripps included the footage in a television report and posted that to YouTube with Content ID protections. Google's software rovers quickly matched the NASA footage as apparently infringing Scripps and alerted Scripps, which had NASA's footage taken down.[9] Not that Scripps really thought NASA had a camera crew on Mars—it acted automatically. Scripps quickly rectified matters with NASA. Limited harm resulted, but perhaps others without rockets at their disposal might not get errors fixed so quickly.

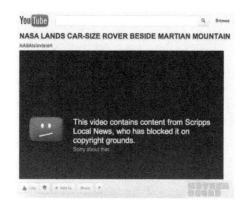

Scripps Mistakenly Claimed Copyright on NASA's Rover Footage

Some criticize systems like Content ID for bypassing the safeguards of the take-down law. Because no take-down notices are sent, copyright owners face no risk for bad faith use of Content ID, nor does YouTube lose immunity for failing to respond to take-down notices. Content ID makes mistakes. It's a little like subjecting public expression to the whims of AutoCorrect. A common glitch is for Content ID to mistakenly identify bird songs as performances of copyrighted music.

Users whose postings are taken down or adorned with advertisements have no legal option. YouTube has no obligation to post user-generated content. But YouTube has made efforts to allow users to respond to Content ID removals.[10] Also, by allowing videos to remain up with ads, the system both encourages postings that may use copyrighted materials and gives copyright owners an incentive to leave the material up. There is something to be said for a flexible system of private agreement and the occasional bit of public outcry. Going to court may be no better. It is expensive, and judges may mistake bird songs for Lady Gaga as well.

Google responded to concerns about Content ID in 2012. Now, when a video is taken down by Content ID at the request of a copyright owner, the poster may request it be put back. For the copyright owner to take it down again, the copyright owner must file a formal Digital Millennium Copyright Act take-down notice, meaning penalties apply if the take-down notice is filed in bad faith.

Copy protection (a.k.a. digital rights management systems, or DRM)

> **copy protection: n.** A class of methods for preventing incompetent pirates from stealing software and legitimate customers from using it. Considered silly.
>
> *The New Hacker's Dictionary*[11]

Copy-protection technology has a long history. Here are a few copy-protection methods for software and games, gleaned from *Wacky copy protection methods from the good old days*:[12]

- *Lenslok*: the game would describe a scrambled code, decipherable only by looking through the plastic Lenslok.

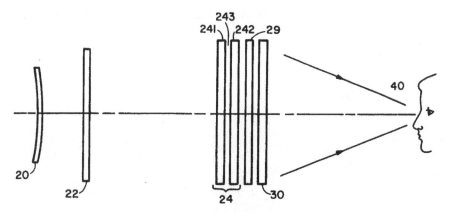

U.S. Statutory Invention Registration H423, by U.S. Navy: "Fresnel Lens in an Improved Infinity Image Display System"

- *Code wheels*: kind of like the Enigma machine, code wheels would yield numbers necessary to play the game.

United States Patent [19]

Wood

[11] **4,154,006**

[45] **May 15, 1979**

[54] CODING AND DECODING MACHINE

[75] Inventor: Chester C. Wood, Washington, D.C.

[73] Assignee: The United States of America as represented by the Secretary of the Navy, Washington, D.C.

[21] Appl. No.: 65,741

[22] Filed: Feb. 25, 1936

[51] Int. Cl.^{2} H04L 9/00
[52] U.S. Cl. 35/3; 340/350; 340/357
[58] Field of Search 35/2, 3, 4; 340/350, 340/354, 357, 358

[56]   **References Cited**
          U.S. PATENT DOCUMENTS

1,984,599	12/1934	Safar	35/6
2,089,603	8/1937	Hagelin	35/3
2,139,676	12/1938	Friedman	35/4

Primary Examiner—Howard A. Birmiel
Attorney, Agent, or Firm—John R. Utermohle

EXEMPLARY CLAIM

3. A coding and decoding machine, comprising a plurality of keys each of which bears a symbol, a plurality of indicating elements each of which represents a symbol on one of said keys, and means to connect each of said keys at random to an indicating element, said means including devices for forming randomly permuting electric circuits between said keys and said indicating elements, operating mechanisms acting at respectively different rates upon different ones of said devices to change said circuits, common driving means for all of said mechanisms, and means for simultaneously disconnecting all of said devices from said devices to permit of setting said devices to desired relative initial positions.

10 Claims, 17 Drawing Figures

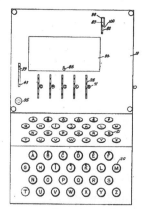

Patent Application Described as "A Plurality of Co-Axially Mounted Independently Rotatable Code Wheels" (Filed 1936, Issued 1979 Due to a Secrecy Order (or a Slow Examiner))

- *Manuals*: the game would require inputting, say, the third word on page five.

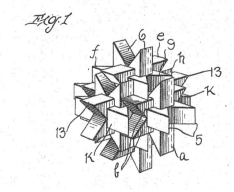

A. P. DULGEROFF.
PUZZLE.
APPLICATION FILED MAY 9, 1914.

1,129,281.

Patented Feb. 23, 1915.
2 SHEETS—SHEET 1.

Fig.1

Puzzles Can Be Used for Copy Protection

- *Dongles*: to use the software requires sticking a special device into one of the computer's ports. This is still in use for many software applications.

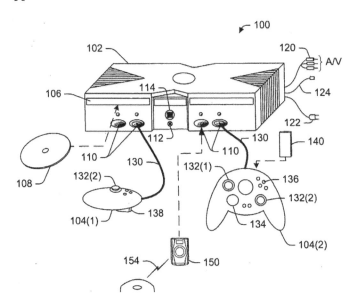

Patent Application from Microsoft for "DVD-Enabling Dongle for a Console-Based Gaming System"

- *Feelies*: thingies needed for the game. In the words of pingdom, "the Hitchhiker's Guide to the Galaxy [game] for the Commodore 64 came with things like 'Peril Sensitive Sunglasses' (opaque black glasses), the order to destroy Earth, Pocket Fluff, a Don't Panic! Button, and a Microscopic Space Fleet (an empty plastic bag)."[13]

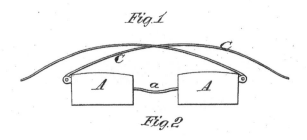

F. YEISER.

SPECTACLES.

No. 170,795.

Patented Dec. 7, 1875.

Fig.1

Fig.2

Spectacles

Copy protection is still with us. However, in some areas, such as music and software, most commercial works do not have copy protection. Copyright owners have learned that anti-copying technology often does little to protect against copying, and it annoys customers. Other areas still have anti-copying technology, such as some e-books and some expensive software packages, which may even require use of a dongle to gain access. People have already developed digital rights management systems for 3D printers.[14]

Copyright law gives legal protection to copy protection technology, making it illegal to circumvent technology that controls access to or copying of copyrighted works. Getting around a dongle to use software may be illegal. Blizzard Entertainment, owner of the online game World of Warcraft, was able to shut down MDY Industries, which sold Glider, software that circumvented access controls to allow WoW players to play on autopilot.

xkcd.com, Randall Munroe

Question

What are the advantages for copyright holders of using access and copy controls on music, video, and games? Disadvantages?

Answer

Advantages might include controlling how widely the work is distributed and being able to charge prices for uses. Controls might also prevent adaptations or uses that are against the rules, such as in a game. Disadvantages might include hassles for potential users, which can mean fewer sales and less distribution, plus loss of potential readers/users/customers. Free works circulate more widely, which can bring unforeseen advantages.

Cybersquatting

Racing for the Domain Name Toys.com (Bought by Toys"R"Us for $5.1 Million)?

People leaving Klondike for Nome Sep. 22nd 1899 down Yukon, Date 22 September 1899 (1899-09-22), Source Alaska Digital Archives. Author Per Edward Larss (died 1941) and Joseph Duclos (1863–1917), Wikimedia Commons

When the Internet became widely used, there was a gold rush for domain names. People for the Ethical Treatment of Animals sought peta.org, but it had been snapped up by People Eating Tasty Animals. Trademark holders often found more tech-savvy parties had bought their domain.com and were willing to hand it over, for a price. Some saw this as simply efficient. The first party to act gets property rights, and the market reallocates them through transactions. That's how gold mining, oil drilling, and patents work: by staking a claim. But some trademark owners saw it more like squatting, ransom, or extortion.

A trademark owner can try to get a domain handed over through arbitration (cheaper and faster) or court (has more teeth) if the registrant acted in "bad faith." Bad faith would be registering cocacola.com in order to sell it to Coca-Cola, or redirecting cocacola.com to a competitor or a click farm. But registering cocacolacritic.com in order to publicize health aspects of soft drinks would not be bad faith. People Eating Tasty Animals registered peta .org in bad faith because its goal was to sell the domain to People for the Ethical Treatment of Animals.

People Eating Tasty Animals

People Eating Tasty Animals is in no way connected with, or endorsed by, **People for the Ethical Treatment of Animals**. Individuals and organizations whose names appear here do not necessarily endorse the contents of this page.

A resource for those who enjoy eating meat, wearing fur and leather, hunting, and the fruits of scientific research (more!).

| Meat | Leather | Fur | Taxidermy | Hunting | Fishing |
| Pets | Research | Miscellaneous | Comments | Hate Mail |

DOMAIN NAME **Domain Name Rights Coalition**
Working to obtain equitable, consistent and responsible domain name policies.

New in this edition: (June 9, 1996)

- Twenty more pieces of "hate mail"
  Mail section fixed to support all browsers
- This site is not a "front!"

PETA?

Mark owners often succeed against typosquatters, who send imperfect typists to pop-up advertisement confinement (such as joecartoon.com getting the handover of joecarton.com). Some disputes are between competing trademarks, as when the World Wildlife Fund (WWF) grappled wwf.com from the World Wrestling Federation (WWF).[15] But Broadcom could not get cyberbroadcom.com, because there were legitimate reasons to use the "broadcom" in its descriptive sense. Google was unable to wrest away oogle.com. The owner had registered in 1999, when he was but 13 years old. The panel did not find evidence of bad faith, even though he offered to sell it to Google for a hefty price.[16]

Questions

1. Instead of buying toys.com for $5.1 million, why didn't Toys"R"Us get a handover order?

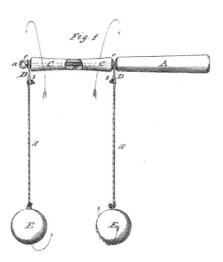

A. R. BATCHELDER.
Toys.
No. 139,533. Patented June 3. 1873.

Toys

2. Will armorgames.com, a video game company, get a handover of armourgames.com from a newer Canadian doppelganger with British spelling? Does it matter if armorgames is better known and has been around longer, albeit in a different country?

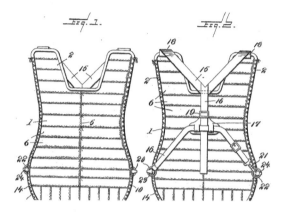

W. J. SULLIVAN.
ARMOR FOR BASE BALL PLAYERS.
APPLICATION FILED MAY 29, 1909.

932,352.

Patented Aug. 24, 1909.

A Different Kind of Sports Armor

Answers

1. A trademark owner can get a handover only for bad faith use of the mark. "Toys.com" could be used in many good faith ways, without using the Toys"R"Us mark.
2. The Canadian company could use a similar mark, in good faith, in another country. No handover.

Sources

For images, we are indebted especially to Rebecca Tushnet's Georgetown Intellectual Property Teaching Resources Database, to Wikipedia (ditto for many quick references, along with Amazon, as a card catalog), and to Google Patent Search. Our articles in the *Northwestern Journal of Technology and Intellectual Property* were good sources for stories.

Some general sources on copyright, patents, trademarks, and trade secrets that we use often, including some hometown preferences, are listed here:

For law students:

ROBERT P. MERGES, PETER S. MENELL & MARK A. LEMLEY, INTELLECTUAL PROPERTY IN THE NEW TECHNOLOGICAL AGE (2012).
JULIE E. COHEN & LYDIA P. LOREN, COPYRIGHT IN A GLOBAL INFORMATION ECONOMY (3rd ed. 2010).
JANICE M. MUELLER, PATENT LAW (4th ed. 2012).
STEPHEN M. MCJOHN, EXAMPLES & EXPLANATIONS: INTELLECTUAL PROPERTY (4th ed. 2012).

For lawyers:

WILLIAM F. PATRY, PATRY ON COPYRIGHT (2007–2015).
DAVID NIMMER, NIMMER ON COPYRIGHT (2013).
DONALD S. CHISUM, CHISUM ON PATENTS (2003).
J. THOMAS MCCARTHY, MCCARTHY ON TRADEMARKS AND UNFAIR COMPETITION (1996).
ROGER M. MILGRIM, MILGRIM ON TRADE SECRETS (1992).
JAY DRATLER, JR. & STEPHEN M. MCJOHN, INTELLECTUAL PROPERTY LAW: COMMERCIAL, CREATIVE, AND INDUSTRIAL PROPERTY (1991).

For current cases:

PATENTLYO, http://patentlyo.com/.
GROKLAW, http://www.groklaw.net/ (no longer active).
ARS TECHNICA, http://arstechnica.com/.
BOING BOING, http://boingboing.net/.
TECHDIRT, https://www.techdirt.com/.
SSRN, http://www.ssrn.com/en/.
THE IPKAT, http://ipkitten.blogspot.com/.
LABORATORIUM, http://2d.laboratorium.net/.
Eric Goldman, TECHNOLOGY & MARKETING LAW BLOG, http://blog.ericgoldman.org/.
SLASHDOT, http://slashdot.org/.
THE TTABLOG, http://thettablog.blogspot.com/.
MADISONIAN, http://madisonian.net/.
NORTHWESTERN JOURNAL OF TECHNOLOGY AND INTELLECTUAL PROPERTY
SUFFOLK LAW JOURNAL OF HIGH TECHNOLOGY
HARVARD JOURNAL OF LAW & TECHNOLOGY
For academics, the IPProf and Cyberprof listservs.

Chapter 1: Copyright: What, Why, When, Whence?

1. *Copyright and Permissions*, MARK TWAIN PROJECT, http://www.marktwainproject.org/copyright.shtml.

2. JOHN TEHRANIAN, INFRINGEMENT NATION: COPYRIGHT 2.0 AND YOU (2011).

3. Oliver Goodenough, *Information Replication in Culture: Three Modes for the Transmission of Culture Elements through Observed Action*, *in* IMITATION IN ANIMALS AND ARTIFACTS 573 (K. Dautenhahn & C. L. Nehaniv eds., 2002).

4. For differences between the UK and U.S. versions, see http://www.hp-lexicon.org/about/books/ps/differences-ps .html.

5. LAWRENCE LESSIG, REMIX: MAKING ART AND COMMERCE THRIVE IN THE HYBRID ECONOMY (2008).

6. DAVID BELLOW, IS THAT A FISH IN YOUR EAR? 137 (2011).

7. *Novelization of a Movie That Was Based on a Book*, THE STRAIGHT DOPE (Sept. 15, 2011, 3:27 PM), http://boards.straightdope.com/sdmb/showthread.php?t=624378.

8. This was once quoted at http://artspiegelmanandmaus.wordpress.com/2011/01/31/maus-the-movie/, though the authors have deleted the site.

9. Ben Sisario, *Site to Resell Music Files Has Critics*, N.Y. TIMES, Nov. 14, 2011, http://www.nytimes.com/2011 /11/15/business/media/reselling-of-music-files-is-contested.html.

10. Michael B. Farrell, *Digital Used Record Store Faces Lawsuit: Case Involving Cambridge's ReDigi Could End with a Landmark Ruling*, BOS. GLOBE, July 2, 2012, https://www.bostonglobe.com/business/2012/07/01/the-used -record-store-goes-digital-music-resale-brings-digital-showdown/vOhr7pzVNiWc2gRjKa9EnN/story.html.

11. Copyright Act, 17 U.S.C. § 106 (2015).

12. Mass. Museum of Contemporary Art Found., Inc. v. Büchel, 593 F.3d 38 (1st Cir. 2010).

13. Wendy J. Gordon, *Intellectual Property*, *in* THE OXFORD HANDBOOK OF LEGAL STUDIES 617 (Peter Cane & Mark Tushnet eds., 2003); DAN HUNTER, THE OXFORD INTRODUCTIONS TO U.S. LAW: INTELLECTUAL PROPERTY (2012).

14. U.S. CONST. art. I, § 8, cl. 8.

15. Campbell v. Acuff-Rose Music, Inc., 510 U.S. 569, 584 (1994).

16. JESSICA SILBEY, HARVESTING INTELLECTUAL PROPERTY: IP INTERVENTIONS AND THE ROLES OF INTELLECTUAL PROPERTIES IN CREATIVE AND INNOVATIVE WORK (2013). Regarding other spurs to creativity, see Bruce Bower, *Vodka Delivers Shot of Creativity: A Boozy Glow May Trigger Problem-Solving Insights*, 181 SCIENCE NEWS, no. 6, at 12 (2012).

17. GOOD COPY, BAD COPY (dir. Andreas Johnsen, Ralf Christensen & Henrik Moltke 2007).

18. Heather Pringle, *The Origins of Creativity*, SCIENTIFIC AM., Feb. 2013, http://www.scientificamerican.com /article/the-origins-of-creativity-creativity-special/.

19. Or you could also consult *The Books of the Century*, http://www.ocf.berkeley.edu/~immer/books1900s.

20. W.S. Gilbert & Arthur Sullivan, THE PIRATES OF PENZANCE (1879):

   FRED. Well, then, it is my duty, as a pirate, to tell you that you are too tender-hearted. For instance, you make a point of never attacking a weaker party than yourselves, and when you attack a stronger party you invariably get thrashed.
   KING. There is some truth in that.
   FRED. Then, again, you make a point of never attacking an orphan!
   SAM. Of course: we are orphans ourselves, and know what it is.
   FRED. Yes, but it has got about, and what is the consequence? Every one we capture says he's an orphan. The last three ships we took proved to be manned entirely by orphans, and so we had to let them go. One would think that Great Britain's mercantile navy was recruited solely from her orphan asylums, which we know is not the case.

For further reading:

A comprehensive survey of styles of remixing was held at the Museum of the Moving Image, in the exhibition *Cut Up: Featuring Supercuts, Mashups, Remixes, and Other Re-Edited Popular Media* (through October 14, 2013).

Chapter 2: What Is Copyrighted: A Creative Work of Authorship in Tangible Form

1. Rebecca Curtin, *The "Capricious Privilege": Rethinking the Origins of Copyright under the Tudor Regime*, 59 J. COPYRIGHT SOC'Y U.S.A. 391 (2012).

2. For a critique of copyright law's concept of the author, see MARTHA WOODMANSEE & PETER JASZI, THE CONSTRUCTION OF AUTHORSHIP: TEXTUAL APPROPRIATION IN LAW AND LITERATURE (1994).

3. 17 U.S.C. § 102(a) (2015). Compilations and derivative works are also included in copyrightable subject matter. 17 U.S.C. § 103(a) (2015).

4. Rebecca Tushnet, *Worth a Thousand Words: The Images of Copyright*, 125 HARV. L. REV. 683 (2012).

5. Mark Frauenfelder, *Zero Ohm Resistor*, BOING BOING (July 6, 2012, 4:05 PM), http://boingboing .net/2012/07/06/0-ohm-resistor.html.

6. Richard M. Stallman, *Did You Say "Intellectual Property"? It's a Seductive Mirage*, GNU.ORG, http://www.gnu .org/philosophy/not-ipr.html.

7. From Copyright Office regulations at *Code of Federal Regulations: Title 37—Patents, Trademarks, and Copyrights*, COPYRIGHT.GOV, http://copyright.gov/title37/:

   § 202.1 Material not subject to copyright.
   The following are examples of works not subject to copyright and applications for registration of such works cannot be entertained:
   (a) Words and short phrases such as names, titles, and slogans; familiar symbols or designs; mere variations of typographic ornamentation, lettering or coloring; mere listing of ingredients or contents; . . .

8. Ralph Clifford, *Intellectual Property in the Era of the Creative Computer Program: Will the True Creator Please Stand Up?*, 71 TUL. L. REV. 1675 (1997); *see* NICK MONTFORT, INTERACTIVE FICTION: OTHER POETIC AND IMAGINATIVE WRITING FOR THE COMPUTER AND WRITING ON DIGITAL MEDIA TOPICS, http://nickm.com /if/#about.

9. James Gleick, *Auto Crrect Ths!*, N.Y. TIMES, Aug. 4, 2012, http://www.nytimes.com/2012/08/05/opinion/sunday /auto-correct-this.html.

10. Mike Masnick, *Monkey Business: Can a Monkey License Its Copyrights to a News Agency?*, TECHDIRT (July 7, 2011), https://www.techdirt.com/articles/20110706/00200314983/monkey-business-can-monkey-license-its -copyrights-to-news-agency.shtml.

11. Robert Beckhusen, *Submarine Commander Faked Own Death to Escape Lover*, WIRED (Sept. 20, 2012), http:// www.wired.com/2012/09/navy-commander/.

12. Cory Doctorow, *Delightful Science Fiction Story in Review of $6800 Speaker Cable*, BOING BOING (Nov. 27, 2010, 10:08 PM), http://boingboing.net/2010/11/27/delightful-science-f.html.
    Other customer reviews of the K2: "I caught a Whale with them"; "Thanks to the AudioQuest K2 Terminated speaker cables, I have become the DEVOURER OF WORLDS!"; "This cable kept the warp drive functional even through the time travel slingshot maneuver."

13. BILL BRYSON, MADE IN AMERICA (2001) (quoting THE ECONOMIST 260).

14. Arthur Lubow, *Can Modern Dance Be Preserved?*, N.Y. TIMES, Nov. 5, 2009, http://www.nytimes.com/2009/11 /08/magazine/08cunningham-t.html?pagewanted=all.

Chapter 3: Movie Rights, Book Deals, Recording Contracts, Free Software: Copyright Transactions

1. For some commissioned works, they may even agree the work is deemed a work made for hire.

2. Ben Kuchera, *Reasons to Leave: Studios Often Own Games Devs Create in Spare Time*, Ars Technica (Sept. 9, 2011), http://arstechnica.com/gaming/2011/09/more-reasons-to-leave-studios-often-own-games-devs-create-in-spare-time/.

3. Jesse Green, *Theater; Edward Albee Returns to the Zoo*, N.Y. Times, May 16, 2004, *available at* http://www.nytimes.com/2004/05/16/theater/theater-edward-albee-returns-to-the-zoo.html.

4. Michael Cieply, *Sony Is Said to Buy Film Rights to Jobs Biography*, Media Decoder (Oct. 8, 2011, 3:34 PM), http://mediadecoder.blogs.nytimes.com/2011/10/08/sony-is-said-to-buy-film-rights-to-jobs-biography/.

5. Mike Masnick, *Why Does Copyright Last 70 Years after Death . . . But Licenses Expire at Death?*, Techdirt (Sept. 4, 2012, 9:15 AM), https://www.techdirt.com/articles/20120904/01275120261/why-does-copyright-last-70-years-after-death-licenses-expire-death.shtml (quoting Kevin Marks).

6. For information on free software, see *GNU General Public License*, GNU.org, http://www.gnu.org/copyleft/gpl.html (last visited June 2, 2015); Jonathan Zittrain, *Normative Principles for Evaluating Free and Proprietary Software*, 71 U. Chi. L. Rev. 265–87 (2004) (describing history of disputes involving rights to Linux, an open source operating system); Stephen McJohn, *The Paradoxes of Free Software*, 9 Geo. Mason L. Rev. 25 (2000); Robert W. Gomulkiewicz, *General Public License 3.0: Hacking the Free Software Movement's Constitution*, 42 Hous. L. Rev. 1015 (2005); Greg Vetter, *Exit and Voice in Free and Open Source Software Licensing: Moderating the Rein over Software Users*, 85 Or. L. Rev. 183 (2006); Jacobsen v. Katzer, 535 F.3d 1373 (Fed. Cir. 2008) (enforcing open source license).

7. *See Why Should I Use the GNU GPL Rather Than Other Free Software Licenses?*, GNU.org, http://www.gnu.org/licenses/gpl-faq.html#WhyUseGPL (last visited June 2, 2015).

8. Jason Toynbee, *Reggae Open Source: How the Absence of Copyright Enabled the Emergence of Popular Music in Jamaica*, in Copyright and Piracy: An Interdisciplinary Critique 357 (Bently, Davis & Ginsburg eds., 2010).

9. *See* Catherine L. Fisk, *Credit Where It's Due: The Law and Norms of Attribution*, 95 Geo. L.J. 49, 90 (2006) ("Attribution is foundational to the modern economy. The reputation we develop for the work we do proves to the world the nature of our human capital. Credit is instrumentally beneficial in establishing a reputation and intrinsically valuable simply for the pleasure of being acknowledged. Indeed, credit is itself a form of human capital.").

10. *License Your Work*, Creative Commons, http://creativecommons.org/choose/ (last visited June 2, 2015); Jacobsen v. Katzer, 535 F.3d 1373 (Fed. Cir. 2008) (holding FOSS license enforceable); Lawrence Lessig, Free Culture: The Nature and Future of Creativity (2005); Michael J. Madison, Brett M. Frischmann & Katherine J. Strandburg, *Constructing Commons in the Cultural Environment*, 95 Cornell L. Rev. 657 (2010); Lawrence B. Solum, *Questioning Cultural Commons*, 95 Cornell L. Rev. 817 (2010); Eric Johnson, *Rethinking Sharing Licenses for the Entertainment Media*, 26 Cardozo Arts & Ent. L.J. 391 (2008).

Chapter 4: Ideas Are Not Copyrighted

1. Litchfield v. Spielberg, 736 F.2d 1352 (9th Cir. 1984).

2. Christopher Sprigman & Dotan Oliar, *There's No Free Laugh (Anymore): The Emergence of Intellectual Property Norms and the Transformation of Stand-Up Comedy*, 94 Va. L. Rev. 1787 (2008); *see also Chloë Sevigny Approved for Second Umlaut*, TheOnion.com (Sept. 12, 2012), http://www.theonion.com/graphic/chloe-sevigny-approved-for-second-umlaut-29485.

3. Capcom Co., Ltd. v. The MKR Group, Inc., 2008 WL 4661479 (N.D. Cal. 2008).

4. Eric Goldman, *Irrational Copyright Lawsuit over "Pi Symphony" Gets Sliced—Erickson v. Blake*, Technology & Marketing Law Blog (March 19, 2012), http://blog.ericgoldman.org/archives/2012/03/putting_the_num.htm (calling the case "an instant copyright classic").

5. Detective Comics v. Bruns Publications, 111 F.2d 432 (2d Cir. 1940).

6. Warner Bros. Inc. v. American Broadcasting Co., 720 F.2d 231 (2d Cir. 1983).

7. Paul Hseih, *Bad Logic*, GEEKPRESS (July 27, 2012, 12:03 AM), http://blog.geekpress.com/2012/07/bad-logic.html.

8. Allen v. Scholastic Inc., 739 F. Supp. 2d 642 (S.D.N.Y. 2011).

9. Nancy Prager, *Good Poets Borrow, Great Poets Steal . . .* , PROTECT AND LEVERAGE (May 8, 2007), http://nancyprager.wordpress.com/2007/05/08/good-poets-borrow-great-poets-steal/.

10. Kaplan v. Stock Market Photo Agency, 133 F. Supp. 2d 317 (S.D.N.Y. 2001).

11. Brandir International v. Cascade Pacific Lumber, 834 F.2d 1142 (2d Cir. 1985).

12. Jonathan H. Liu, *The 5 Best Toys of All Time*, WIRED (Jan. 31, 2011, 8:00 AM), http://archive.wired.com/geekdad/2011/01/the-5-best-toys-of-all-time/all/.

13. For sources on copyright and software, see Pamela Samuelson, Randall Davis, Mitchell D. Kapor & J.H. Reichman, *A Manifesto Concerning the Legal Protection of Computer Programs*, 94 COLUM. L. REV. 2308 (1994) (keenly describing code as language that behaves); Marci A. Hamilton & Ted Sabety, *Computer Science Concepts in Copyright Cases: The Path* to a *Coherent Law*, 10 HARV. J.L. & TECH. 239 (1997).

14. Tetris Holding, LLC v. Xio Interactive, Inc., 863 F. Supp. 2d 394 (D.N.J. 2012). For further reading, see Greg Lastowka, *Spry Fox Attacks the Clones: Is Palpatine Behind This?*, TERRA NOVA (Sept. 28, 2012), http://terranova.blogs.com/terra_nova/2012/09/a-few-random-thoughts-on-triple-town.html.

15. Rob Beschizza, *Early iPhone Mockups Show Sony Influence*, BOING BOING (July 27, 2012, 8:17 AM), http://boingboing.net/2012/07/27/early-iphone-mockups-show-sony.html.

16. Apple Inc. v. Samsung Elecs. Co., No. 2014-1335 (Fed. Cir. May 18, 2015). In the United States, an appeals court held that Apple infringed patents of Samsung but that the look of the iPhone was not protected by trademark, meaning that a $930 million jury verdict for Apple was sent down to be recalculated.

17. Chris Foresman, *UK Judge: Galaxy Tab Not "Cool" Enough to Infringe iPad Design*, ARS TECHNICA (July 9, 2012, 1:10 PM), http://arstechnica.com/apple/2012/07/uk-judge-galaxy-tab-not-cool-enough-to-infringe-ipad-design/ (quoting Bloomberg News report).

18. The site PatentlyO.com provides much information about the history and growing importance of design patents.

Chapter 5: Formalities: For Want of a ©, the Kingdom Was Lost

1. *Robin Williams: 50 Great Quotes*, THE TELEGRAPH, http://www.telegraph.co.uk/culture/culturepicturegalleries/11027564/Robin-Williams-50-great-quotes.html?frame=3003238 (last visited July 26, 2015) ("In England, if you commit a crime, the police don't have a gun and you don't have a gun. If you commit a crime, the police will say, 'Stop, or I'll say stop again'").

Chapter 6: Fair Use: Excelsior!

1. William Patry literally wrote the book on THE FAIR USE PRIVILEGE IN COPYRIGHT LAW (2d ed. 2012). Two comprehensive studies of fair use case law are Pamela Samuelson, *Unbundling Fair Uses*, 77 FORDHAM L. REV. 2537 (2009), and Barton Beebe, *An Empirical Study of U.S. Copyright Fair Use Opinions, 1978–2005*, 156 U. PA. L. REV. 549 (2008).

2. Alfred C. Yen, *A First Amendment Perspective on the Idea/Expression Dichotomy and Copyright in a Work's Total Concept and Feel*, 38 EMORY L.J. 393 (1989).

3. *See* Eldred v. Ashcroft, 537 U.S. 186, 219–20 (2003).

4. And lack of copyright can spur innovators to act quickly. Kal Raustiala & Chris Sprigman, *The Piracy Paradox: Innovation and Intellectual Property in Fashion Design*, 92 VA. L. REV. 1687 (2006).

5. Joseph P. Liu, *Two-Factor Fair Use?*, 31 COLUM. J.L. & ARTS 4 (2008).

6. Laura A. Heymann, *Everything Is Transformative: Fair Use and Reader Response*, 31 Colum. J.L. & Arts 445 (2008).

7. Leibovitz v. Paramount Pictures Corp., 137 F.3d 109, 111 (2d Cir. 1998).

8. James Gibson, *Risk Aversion and Rights Accretion in Intellectual Property Law*, 116 Yale L.J. 882 (2007).

9. Gaylord v. United States, 85 Fed. Cl. 59, 62 (2008).

10. Association for Renaissance Martial Arts, *Samples of Medieval & Renaissance Unarmed Fighting Skills*, TheARMA.org, http://www.thearma.org/spotlight/unarmedcombat.htm#.VcvBznFVikp.

11. Iowa State Univ. Research Found., Inc. v. Am. Broad. Co., 621 F.2d 57, 60 (2nd Cir. 1980).

12. Matthew Sag, *Copyright and Copy-Reliant Technology*, 103 Nw. U. L. Rev. 1607 (2009).

For further reading:

For three articles on how fair use is intersecting with fast-changing technologies, see Edward Lee, *Technological Fair Use*, 83 S. Cal. L. Rev. 797, 817 (2010); Fred von Lohmann, *Fair Use as Innovation Policy*, 23 Berkeley Tech. L.J. 829 (2008); and Randal C. Picker, *Fair Use v. Fair Access*, 31 Colum. J.L. & Arts 603 (2008).

Chapter 7: Copyright Lawsuits

1. Selle v. Gibb, 741 F.2d 896, 900 (7th Cir. 1984).

2. *It's about Time*, https://www.medievalscribes.com/.

3. Robin Feldman, *The Role of the Subconscious in Intellectual Property Law*, 2 Hastings Sci. & Tech. L.J. 1 (2010).

4. Karen Sloan, *James Joyce Estate Agrees to Pay Plaintiff's Fees in Fair Use Dispute*, Nat'l L.J. (Sept. 30, 2009), http://www.law.com/jsp/article.jsp?id=1202434181383.

5. Bouchat v. Baltimore Ravens Ltd. P'ship, 619 F.3d 301 (4th Cir. Sept. 2, 2010) (order to blur logo on films); Mark A. Fischer & Meg Sallay, *Nevermore the Ravens' History?*, Law360 (Feb. 4, 2011), http://www.law360.com/articles/223333/nevermore-the-ravens-history.

Chapter 8: Patentable Inventions: Products and Processes

1. The Emerson saying is itself an improvement. His original words: "If a man has good corn or wood, or boards, or pigs, to sell, or can make better chairs or knives, crucibles or church organs, than anybody else, you will find a broad hard-beaten road to his house, though it be in the woods."

2. Letter from Thomas Jefferson to Isaac McPherson (Aug. 13, 1813), http://www.monticello.org/site/research-and-collections/patents.

3. *Id.*

4. B. Zorina Khan, The Democratization of Invention: Patents and Copyrights in American Economic Development, 1790–1920, at 214–15 (2005).

5. *Monty Python's Flying Circus: Hungarian Phrasebook Sketch (Episode 25)* (1970).

6. Bill Bryson, At Home 397 (2010).

7. *Interesting Patents*, Gallagher & Dawsey, http://www.invention-protection.com/ip/inventor_resources/interesting_patents.html (last visited June 3, 2015).

8. Donald S. Chisum & Michael A. Jacobs, Understanding Intellectual Property Law 2–19 (1992).

9. U.S. Patent No. X000001 (filed 1790 by Samuel Hopkins) (the X patents were lost).

10. U.S. Patent No. 2,286,644, Method and Apparatus for Processing Potatoes (filed 1937).

11. Press Release, Electronic Frontier Foundation, Movie Legend Hedy Lamarr to Be Given Special Award at EFF's Sixth Annual Pioneer Awards (Mar. 11, 1997), http://w2.eff.org/awards/pioneer/1997.php.

12. U.S. Patent 4,683,195, Process for Amplifying, Detecting, and/or Cloning Nucleic Acid Sequences (filed 1987 by Kary B. Mullis et al.).

13. Jim Bieberich, *Historical U.S. Patent Collection*, USPAT.COM, http://www.uspat.com/historical/ (last visited June 3, 2015); *Nobel Prize Winners: Contributions to Electricity, Electronics, Electromagnetism, and Electrochemistry*, ELECTRO.PATENT-INVENT.COM, http://www.electro.patent-invent.com/electricity/nobelwinners.html (last visited June 3, 2015); LEMELSON-MIT.EDU, http://lemelson.mit.edu/ (last visited Aug. 17, 2015).

14. BILL BRYSON, AT HOME 237 (2010).

15. Dennis Crouch, *Tees & Radar: Patent Pendency 2011*, PATENTLYO.COM (July 20, 2011), http://patentlyo.com/patent/2011/07/tees-radar-patent-pendency-2011.html.

Chapter 9: Nonpatentable: Humans, Nature, Ideas

1. Ananda M. Chakrabarty, Diamond v. Chakrabarty: *A Historical Perspective, in* DONALD CHISUM ET AL., PRINCIPLES OF PATENT LAW 758 (2d ed. 2001).

2. Ker Than, *Neanderthals, Humans Interbred—First Solid DNA Evidence*, NAT'L GEOGRAPHIC (May 8, 2010), http://news.nationalgeographic.com/news/2010/05/100506-science-neanderthals-humans-mated-interbred-dna-gene/ ("According to a new DNA study, most humans have a little Neanderthal in them—at least 1 to 4 percent of a person's genetic makeup. The study uncovered the first solid genetic evidence that 'modern' humans—or *Homo sapiens*—interbred with their Neanderthal neighbors, who mysteriously died out about 30,000 years ago.").

3. Even if the adrenaline occurs in nature, the purified form is for "every practical purpose a new thing commercially and therapeutically." Parke-Davis & Co. v. H.K. Mulford Co., 189 F. 95 (C.C.S.D.N.Y. 1911).

4. Ass'n for Molecular Pathology v. Myriad Genetics, Inc., 133 S. Ct. 2107 (2013).

5. Gina Kolata, *In Preventing Alzheimer's, Mutation May Aid Drug Quest*, N.Y. TIMES, July 11, 2012, http://www.nytimes.com/2012/07/12/health/research/rare-gene-mutation-is-found-to-stave-off-alzheimers.html.

6. Velvet Type Fabric and Method of Producing Same, U.S. Patent No. 2,717,437 (filed Oct. 15, 1952).

7. Gottschalk v. Benson, 409 U.S. 63 (1972).

8. Bilski v. Kappos, 561 U.S. 593 (2010).

9. Diamond v. Diehr, 450 U.S. 175 (1981).

10. Catherine Rampell, *2 from U.S. Win Nobel in Economics*, N.Y. TIMES, Oct. 15, 2012, http://www.nytimes.com/2012/10/16/business/economy/alvin-roth-and-lloyd-shapley-win-nobel-in-economic-science.html.

11. Tracy Tullis, *How Game Theory Helped Improve New York City's High School Application Process*, N.Y. TIMES, Dec. 5, 2014, http://www.nytimes.com/2014/12/07/nyregion/how-game-theory-helped-improve-new-york-city-high-school-application-process.html.

Chapter 10: New, Useful, and Nonobvious

1. BILL BRYSON, MADE IN AMERICA 92 (1994).

2. *Id*. at 87–89.

3. Leader Technologies Inc. v. Facebook Inc., No. 11-1366 (Fed. Cir. May 8, 2012).

4. Cory Doctorow, *Did Donald Duck Foil a Patent Application?*, BOING BOING (Sept. 26, 2011, 7:41 AM), http://boingboing.net/2007/09/26/did-donald-duck-foil.html.

5. *In re* Montgomery, No. 11-1376 (Fed. Cir. May 8, 2012).

6. *Léonard de Vinci: Projets, Dessins, Machines* (exhibit at Cité des Sciences et de l'Industrie, Paris, France, 2013).

7. Honeywell v. Sperry Rand Corp., 180 U.S.P.Q. (BNA) 673 (D. Minn. 1973).

8. Michael Risch, *Utility: A Surprisingly Useful Requirement*, 19 Geo. Mason L. Rev. 57 (2011).

9. *See, e.g.*, *In re* Fisher, 421 F.3d 1365 (Fed. Cir. 2005).

10. Karen Rowan, *Top 10 Ig Nobels: The Best of Science's Strangest Prize*, Popular Mechs. (Sept. 30, 2009), http://www.popularmechanics.com/science/4306064.

11. David Pescovitz, *Antigravity Device Patented*, Boing Boing (Nov. 17, 2011, 2:59 PM), http://boingboing .net/2005/11/17/antigravity-device-p.html (quoting Philip Ball, *Antigravity Craft Slips Past Patent Officers*, 438 Nature 139 (2005)).

12. Reckendorfer v. Faber, 92 U.S. 347 (1875).

13. Method of Treating Foodstuff, U.S. Patent No. 2,495,429 (filed Oct. 8, 1945).

14. For more on accidental inventions, see Sean B. Seymore, *Serendipity*, 88 N.C. L. Rev. 187–211 (2009).

15. Crocs, Inc. v. U.S. International Trade Commission, 598 F.3d 1294 (Fed. Cir. 2010).

16. Western Union Co. v. MoneyGram Payment Systems Inc., 626 F.3d 1361 (Fed. Cir. 2011).

Chapter 11: Getting a Patent

1. For a list of patent classifications, see *Class Schedule for Class 74*, U.S. Patent & Trademark Office, http://www.uspto.gov/web/patents/classification/uspc074/sched074.htm (last visited July 26, 2015).

2. Dan L. Burk & Mark A. Lemley, The Patent Crisis and How the Courts Can Solve It 22–26 (2009).

3. Janice Mueller, Patent Law (4th ed. 2012) (quoting Judge Giles Rich, 67).

4. *See* Robert C. Faber, *Claim Forms and Formats in General* (ch. 2), *in* Faber on Mechanics of Patent Claim Drafting (2008).

5. Matt Lowrie, Suffolk Law School, Presentation at Suffolk University Law School Symposium: The CSI Effect: Litigation Strategies and Courtroom Dynamics (May 10, 2007).

6. Thanks to Joseph Koipally at Fish & Richardson and his colleagues for some bespoke anachronistic claims:

Drafter I

1. An article of manufacture, comprising a disc that defines a continuous or substantially continuous circular outer periphery.

2. The article of manufacture of claim 1, wherein the disc defines opposed faces inside the outer periphery and a central passage that extends at least partially between the opposed faces, and further comprising a central column in the central passage.

3. The article of manufacture of claim 2, wherein the central column is fixedly attached to the disc so as to rotate in unison with the disc.

4. The article of manufacture of claim 3, wherein the central column extends outward from each of the opposed faces a sufficient distance so that a human may rest a foot on respective sides of the disc on top of the central column (the B.C. wheel).

5. The article of manufacture of claim 1, wherein the disc includes a rim arranged at the outer periphery, a hub near a central rotational axis of the disc, and a plurality of spokes that connect the hub to the rim.

Drafter II

A waggon wheel, including:

(A) a rigid, circular, outer rim;

(B) a cylindrical hub, the hub defining a central channel for receiving an axle, the hub being disposed near a center of the rim; and

(C) a plurality of spokes, the spokes being spaced apart from one another and extending radially from the hub to the rim.

Drafter III

1. A wheel comprising an outer structure defining a closed perimeter.

2. A wheel in accordance with claim 1 wherein the closed perimeter is circular.

3. A wheel in accordance with claims 1 or 2 and further comprising a hub within the outer structure and inter-connecting structure extending between the hub and outer structure.

4. A wheel in accordance with claim 3 wherein the interconnecting structure comprises spokes.

This drafter also added: I use wheel claims to discuss the need for the inventor of the broad claim and inventor of the narrow claim to cross license each other. Broad: hub and rim. Narrow: hub and circular rim.

7. ADRIAN JOHNS, PIRACY 100 (2009) (discussing 18th-century druggists).

8. Atul Gawande, *The Score: How Childbirth Went Industrial*, THE NEW YORKER, Oct. 9, 2006, http://www.newyorker.com/magazine/2006/10/09/the-score.

9. Clarissa Long, *Patent Signals*, 69 U. CHI. L. REV. 625, 636–37 (2002) (describing how patents can serve to signal the strength of a company).

10. THE CAMBRIDGE ECONOMIC HISTORY OF MODERN BRITAIN 122 (Paul A. Johnson & Roderick Floud eds., 2004).

11. TECHNOLOGY IN AMERICA: A HISTORY OF INDIVIDUALS AND IDEAS 66 (Carroll W. Pursell ed., 1990).

12. AUTUMN STANLEY, MOTHERS AND DAUGHTERS OF INVENTION: NOTES FOR A REVISED HISTORY OF TECHNOLOGY (1995).

13. A HAMMER IN THEIR HANDS: A DOCUMENTARY HISTORY OF TECHNOLOGY AND THE AFRICAN-AMERICAN EXPERIENCE (Carroll W. Pursell ed., 2005) (discussing numerous inventions patented by African-American inventors).

14. C. WAYNE SMITH & JOE TOM COTHREN, COTTON: ORIGIN, HISTORY, TECHNOLOGY AND PRODUCTION 78 (1999) (discussing market competition sparked by patented cotton gin "which used spikes attached to a cylinder rather than the saws used in modern gins").

15. CHARLES JOHN PHILLIPS, GLASS: THE MIRACLE MAKER: ITS HISTORY, TECHNOLOGY AND APPLICATIONS 184 (1941).

16. *See, e.g.*, ROBERT S. WOODBURY, HISTORY OF THE GEAR-CUTTING MACHINE: A HISTORICAL STUDY IN GEOMETRY AND MACHINES 106 (1958) ("Another hobbing machine of which we have only the patent was that of Henry Belfield. . . . See his Patent No. 120023 of Oct. 17, 1871."). *See also* Jo Carrillo, *Protecting a Piece of American Folklore: The Example of the Gusset*, 4 J. INTELL. PROP. L. 203, 232 n.138 (1997) ("As there is no patent number or mark on the single remaining 'Ladies' Hiking Tog' garment that survives in the Levi Strauss & Co. Archives, the garment itself confirms Levi historian McDonough's statement that it was not constructed under a patent. But this type of gusset, which is distinct from the public domain gusset, was eventually patented. See U.S. Patent No. 4,392,259; U.S. Patent No. 3,745,589, 'Triangular Crotches for Trousers,' issued to Ebbe Bruno Borsing (Jul. 17, 1973); U.S. Patent No. 478,190.") (cross citations omitted).

17. Randy Kennedy, *A Pollock, in the Eyes of Art and Science*, N.Y. TIMES, Feb. 4, 2007, at A18.

18. U.S. Patent No. 6,469 (filed March 10, 1849) (issued May 22, 1849).

19. Dan Cahoy, Joel Gehman & Zhen Lei, *Fracking Patents: The Emergence of Patents as Information Containment Tools in Shale Drilling*, 19 MICH. TELECOMM. & TECH. L. REV. 279 (2013).

20. Brenda M. Simon, *Patent Cover-Up*, 47 HOUS. L. REV. 1299 (2011), quoted in Cahoy, Gehman, & Lei, *supra* note 19. *See also* Dennis Crouch, *Philips v. AWH Takes a Casualty: "Interface" Construed as "Parallel Bus Interface,"* PATENTLYO (May 11, 2006), www.patentlyo.com/patent/2006/05/philips_v_awh_t.html.

21. Centocor Ortho Biotech, Inc. v. Abbott Labs., 636 F.3d 1341 (Fed. Cir. 2011).

22. *See* Enzo Biochem v. Gen-Probe, 296 F.3d 1316 (Fed. Cir. 2002).

Chapter 12: To Patent or Not to Patent

1. John R. Allison, Mark A. Lemley, Kimberly A. Moore & R. Derek Trunkey, *Valuable Patents*, 92 Geo. L.J. 435 (2004).

2. Bryan Gardiner, *Glass Works: How Corning Created the Ultrathin, Ultrastrong Material of the Future*, WIRED, Sept. 24, 2012, http://www.wired.com/2012/09/ff-corning-gorilla-glass/.

3. *See* Petra Moser, *Patents and Innovation: Evidence from Economic History* (Stanford Law and Econ. Olin Working Paper No. 437, 2012).

4. DAN L. BURK & MARK A. LEMLEY, THE PATENT CRISIS AND HOW THE COURTS CAN SOLVE IT (2009).

5. *Id.*

6. *See* Advertisement for 2008 Mercury Mariner Hybrid, SCIENTIFIC AM., Sept. 2008, at 42.

7. *See* 35 U.S.C. § 287(a) (2010).

8. U.S. Patent No. 6,687,593 (issued February 3, 2004).

9. Robert Farley, *Ad from Sen. Barbara Boxer Attacks Carly Fiorina for Layoffs at HP*, POLITIFACT.COM (Sept. 17, 2010, 4:46 PM), http://www.politifact.com/truth-o-meter/statements/2010/sep/17/barbara-boxer/ad-sen-barbara-boxer-attacks-carly-fiorina-layoffs/.

10. Mark R. Patterson, *When Is Property Intellectual? The Leveraging Problem*, 73 S. CAL. L. REV. 1133 (2000) ("Patents and copyrights protect inventions and expression; they do not protect products.").

11. Control Apparatus, U.S. Patent Application, filed by Claude E. Shannon Aug. 1951 (dropped Jan. 21, 1954).

12. *The Secret Patents for the Atomic Bomb*, ALEXWELLERSTEIN.COM, www.alexwellerstein.com/atomic_patents (last visited June 19, 2015).

13. *See* ADAM B. JAFFE & JOSH LERNER, INNOVATION AND ITS DISCONTENTS: HOW OUR BROKEN PATENT SYSTEM IS ENDANGERING INNOVATION AND PROGRESS AND WHAT TO DO ABOUT IT (2004).

14. Logan Ward et al., *Breakthrough Awards*, POPULAR MECHS., Nov. 2008, at 69, 73.

15. Elizabeth Cooney, *Dueling Emails in MIT Stem Cell Scientist's Tenure Case*, WHITE COAT NOTES (Jan. 29, 2007), http://www.boston.com/yourlife/health/blog/2007/01/mit_provost_pro.html (professor citing his "12 patent applications and technology disclosures").

16. Charles Petit, *Invisibility Uncloaked*, SCI. NEWS, Nov. 6, 2009. In any event, the paper was eventually published in one of the leading journals, *Science*.

17. Andrew Beckerman-Rodau, *The Choice between Patent Protection and Trade Secret Protection: A Legal and Business Decision*, 84 J. PAT. & TRADEMARK OFF. SOC'Y 371 (2002).

18. Kim Bhasin, *Elon Musk: "If We Published Patents, It Would Be Farcical,"* BUS. INSIDER (Nov. 9, 2012, 11:58 AM), http://www.businessinsider.com/elon-musk-patents-2012-11.

19. Marilyn Berger, *Isidor Isaac Rabi, a Pioneer In Atomic Physics, Dies at 89*, N.Y. TIMES, Jan. 12, 1988, http://www.nytimes.com/1988/01/12/obituaries/isidor-isaac-rabi-a-pioneer-in-atomic-physics-dies-at-89.html.

20. *High-Wire Act for Science*, L.A. TIMES, Jan. 9, 2000, at M4; Nicholas Wade, *Scientist at Work: Joe Z. Tsien; Of Smart Mice and an Even Smarter Man*, N.Y. TIMES, Sept. 7, 1999, at F1.

21. Geoff Brumfiel, *Andre Geim: In Praise of Graphene*, NATURE, Oct. 7, 2010, http://www.nature.com/news/2010/101007/full/news.2010.525.html.

Chapter 13: Patent Litigation: The Sport of Kings

1. *See* Phil Swain, Address at Suffolk University Advanced Legal Studies Conference, Current and Future Trends in Patent Law (2007); and Douglas J. Kline, *Patent Litigation: The Sport of Kings*, MIT TECH. REV., Apr. 28, 2004, http://www.technologyreview.com/news/402686/patent-litigation-the-sport-of-kings/.

2. The Crocs claim is edited a bit, for clarity. Here's the full claim.

2. A footwear piece comprising:

a base section including an upper and a sole formed as a single part manufactured from a moldable foam material; and

a strap section formed of a molded foam material attached at opposite ends thereof to the base section such that the strap section is in direct contact with the base section and pivots relative to the base section; and

wherein the upper includes an open rear region defined by an upper opening perimeter; and wherein the sole includes a rear perimeter; and wherein the strap section pivots between a first contact point on the upper opening perimeter and a second contact point on the rear perimeter, and wherein frictional forces developed by the contact between the strap section and the base section at the points of attachment are sufficient to maintain the strap section in place in an intermediary position after pivoting whereby the strap section lends support to the Achilles portion of a human foot inserted in the open rear region; and

wherein the upper includes a substantially horizontal portion and a substantially vertical portion forming a toe region that generally follows the contour of a human foot, wherein the toe region tapers from the inner area of the base section where the larger toes exist to the outer area of the base section where the smaller toes exist; and

wherein a decorative pattern of raised bumps is molded or otherwise created in the upper near to and extending the length of the upper opening perimeter; and

wherein a plurality of ventilators are formed in both the substantially vertical portion and the substantially horizontal portion, and wherein the ventilators extend up a majority of the height of the vertical portion;

wherein the vertical portion of the upper includes an upper strip, wherein the ventilators are formed in the upper strip, and wherein the upper strip extends from the toe region to the points of attachment for the strap section, and wherein the sole includes a lower strip that parallels the upper strip and is separated by a line that extends from the toe region to a heel of the footwear piece, and wherein the lower strip vertically rises in a direction toward the heel; and

wherein the sole includes a bottom surface having front and rear tread patterns longitudinally connected by a flat section without tread patterns bounded by raised side portions; and

wherein the sole further includes a top surface having a support base including a raised pattern where a foot contacts the support base.

3. BILL BRYSON, MADE IN AMERICA 88 (1994).

4. The examples of claim interpretation in the text are drawn from these cases: Ortho-McNeil Pharm., Inc. v. Teva Pharm. Indus., Ltd., 93 U.S.P.Q.2d (BNA) 1125 (Fed. Cir. 2009); Baldwin Graphic Sys., Inc. v. Siebert, Inc., 512 F.3d 1338 (Fed. Cir. 2008); O2 Micro Int'l Ltd. v. Beyond Innovation Tech. Co., 521 F.3d 1351, 1359 (Fed. Cir. 2008); Vehicle IP, LLC v. GMC, 306 F. App'x 574 (Fed. Cir. 2009); Mangosoft, Inc. v. Oracle Corp., 525

F.3d 1327 (Fed. Cir. 2008); Decisioning.com, Inc. v. Federated Dep't Stores, Inc., 527 F.3d 1300 (Fed. Cir. 2008); Welker Bearing Co. v. PHD, Inc., 550 F.3d 1090, 1099 (Fed. Cir. 2008); Ortho-McNeil Pharm., Inc. v. Caraco Pharm. Labs., Ltd., 476 F.3d 1321, 1328 (Fed. Cir. 2007); *see, e.g.*, Envirco Corp. v. Clestra Cleanroom, Inc., 209 F.3d 1360, 1364 (Fed. Cir. 2000) (interpreting whether use of "means" in claim triggered means-plus-function rule); *In re* Buszard, 504 F.3d 1364, 1366 (Fed. Cir. 2007).

5. *See* John Scanlan, *Samuel Johnson's Legal Thought, in* SAMUEL JOHNSON AFTER 300 YEARS 112 (2009).

6. Vehicle IP v. Gen. Motors Corp., No. 2008-1259 (Fed. Cir. 2009).

7. Baran v. Med. Device Techs. Inc., No. 2010-1058 (Fed. Cir. 2010).

8. Power-One Inc. v. Artesyn Techs. Inc., No. 2008-1501 (Fed. Cir. 2010).

9. Edwards Lifesciences LLC v. Cook Inc., No. 2009-1006 (Fed. Cir. 2009).

10. Martek Biosciences Corp. v. Nutrinova, Inc., 579 F.3d 1363, 1382 (Fed. Cir. 2009).

11. To quote Marshall Mathers in "The Real Slim Shady," "We ain't nothing but mammals."

12. *Martek*, 579 F.3d at 1382.

13. Brit. Telecomms. PLC v. Prodigy Commc'ns Corp., 217 F. Supp. 2d 399 (S.D.N.Y. 2002).

14. Mark Lemley, *Rational Ignorance at the Patent Office*, 95 Nw. U. L. REV. 1437 (2001).

15. Christopher A. Cotropia & Mark A. Lemley, *Copying in Patent Law*, 87 N.C. L. REV.1421 (2009). As of 2011, the patent statute does provide prior user rights for someone who used a manufacturing or other commercial process, if they can establish they began using it more than one year before the patent application was filed. 35 U.S.C. § 273.

16. JAMES BESSEN & MICHAEL J. MEURER, PATENT FAILURE: HOW JUDGES, BUREAUCRATS, AND LAWYERS PUT INNOVATORS AT RISK (2008).

Chapter 14: Trademarks

1. SLIM SHADY is a registered trademark of Mathers, Marshall B. III for musical sound recordings and live music performances, according to the Trademark Electronic Search System (TESS), U.S. PATENT AND TRADEMARK OFFICE, http://tmsearch.uspto.gov/bin/gate.exe?f=searchss&state=4802:dtyuuu.1.1.

2. Mark McKenna, *A Consumer Decision-Making Theory of Trademark Law*, 98 VA. L. REV. 67 (2012).

3. Some good sources of marks/logos include INTERBRAND.COM, and *2015 BrandZ Top 100 Global Brands*, MILLWARDBROWN.COM, www.millwardbrown.com/brandz/Top_100_Global_Brands.aspx (last visited June 20, 2015).

4. Andrew Orlowski, *Feds Seize Biker Gang's Trademark*, THE REGISTER (Oct. 22, 2008), www.theregister .co.uk/2008/10/22/doj_seizes_biker_trademark/. The forfeiture order was later reversed on several theories.

5. Qualitex v. Jacobson Prods., 514 U.S. 159, 162–63 (1995).

6. For more logos with hidden meaning, see *20 Best Logos with Hidden Meaning*, TRUEKOLOR (Aug. 11, 2010), www.truekolor.net/20-best-logos-with-hidden-meaning/.

7. *See* Martin Schwimmer, *Some of the Oldest Trademarks in the World*, THE TRADEMARK BLOG (Oct. 2, 2011), http://www.schwimmerlegal.com/2002/10/some-of-the-oldest-trademarks-in-the-world.html.

8. Bayer retained its position as the leading seller despite lack of trademark protection. *See* Glynn S. Lunney, Jr., *Trademark Monopolies*, 48 EMORY L.J. 367, 454 n.284 (1999).

9. RICHARD RHODES, DARK SUN 48 (1995).

10. Casey Johnston, *Google Forces a New Swedish Word Out of (Official) Existence*, ARS TECHNICA (Mar. 26, 2013), http://arstechnica.com/business/2013/03/google-forces-a-new-swedish-word-out-of-official-existence/.

11. Lauren Harrell, *41 Brand Names People Use as Generic Terms*, MENTAL FLOSS, http://www.mentalfloss.com /blogs/archives/93858#ixzz21Y7NXuwh (last visited June 20, 2015).

12. Leah Grinvald, *Shaming Trademark Bullies*, 2011 Wis. L. Rev. 625 (2011).

13. Christina Warren, *Jack Daniel's Sends the Most Polite Cease-and-Desist Letter Ever*, Mashable (July 22, 2012), http://mashable.com/2012/07/22/jack-daniels-trademark-letter/.

Chapter 15: Requirements to Be a Mark

1. Bill Bryson, The Mother Tongue: English and How It Got That Way (2001).

2. *Simpsons* marks, here and elsewhere in the book, are taken from Nancy Friedman's compilation *Simpsons Brand-o-Rama*, Fritinancy (June 29, 2007), http://nancyfriedman.typepad.com/away_with_words/2007/06/simpsons_brando.html.

3. Mark Liberman, *Watson v. Watson*, Language Log (Apr. 21, 2012, 8:32 AM), http://languagelog.ldc.upenn.edu/nll/?p=3918#more-3918.

4. Leo Rosten, The New Joys of Yiddish: Completely Updated 266 (2010).

5. Kenneth L. Port, *On Nontraditional Trademarks* (William Mitchell Legal Studies Research Paper No. 2010-05). "This piece regards nontraditional trademarks like sound, color, scent or even the vertical opening motion of a Lamborghini car door."

6. But courts have held that the shape and color may have lost their trademark function. *See* Shire U.S., Inc. v. Barr Labs., Inc., 329 F.3d 348, 351 (3d Cir. 2003).

Chapter 16: Choosing, Registering, and Owning a Trademark

1. Graham Morrison, *Thanks Linus*, 163 Linux Format 46 (Nov. 2012).

2. Mark Fischer, Intellectual Property Law Developments, Suffolk Law Advanced Legal Studies Conference (2015).

3. A fun article on picking business names: Julian Shapiro, *Before Naming Your Startup, Read This*, TNW News (Apr. 22, 2012, 12:31 PM), http://thenextweb.com/entrepreneur/2012/04/22/before-naming-your-startup-read-this/.

4. John Colapinto, *Famous Names*, The New Yorker, Oct. 3, 2011, http://www.newyorker.com/magazine/2011/10/03/famous-names.

5. *Id.*

6. *Id.*

7. Sonia K. Katyal, *Stealth Marketing and Antibranding: The Love That Dare Not Speak Its Name*, 58 Buff. L. Rev. 795 (2010) (quoted in Stacey Dogan, *Brand or Anti-Brand?*, Intellectual Property Jotwell (Dec. 3, 2010), http://ip.jotwell.com/brand-or-anti-brand/).

Chapter 17: Trademark Infringement: Confusingly Similar?

1. *Brown Out*, Snopes.com, http://www.snopes.com/music/artists/vanhalen.asp (last visited June 20, 2015).

2. On the finer points of commercial use, see Stacey Dogan, *Beyond Trademark Use*, 8 J. Telecomms. & High Tech. L. 135 (2010).

3. Elinor Mills, *A Son Named "Google,"* CNet (Oct. 18, 2005), http://www.cnet.com/news/a-son-named-google/.

4. William James, The Principles of Psychology (1890) (on a baby's impression of the world).

5. JL Beverage Co. v. Beam Inc., No. 00417 (D. Nev. Sept. 25, 2012).

6. Graeme B. Dinwoodie, William O. Hennessey, Shira Perlmutter & Graeme W. Austin, International Intellectual Property Law and Policy 298 (2d ed. 2008).

7. Timothy B. McCormack, *Red with Envy: Can You Trademark a Color?*, Seattle Pi (Nov. 9, 2012), http://blog.seattlepi.com/timothymccormack/2012/11/09/red-with-envy-can-you-trademark-a-color/.

8. Steven Levy, Hackers: Heroes of the Computer Revolution (1984).

9. DANIEL DEFOE, THE COMPLETE ENGLISH TRADESMAN (1726).

10. Hormel Foods Corp. v. Jim Henson Prods., 73 F.3d 497, 503 (2d Cir. 1996).

11. Lena Williams, *Can a Green Candy Make Love Sweeter?*, N.Y. TIMES, Mar. 17, 1993, http://www.nytimes.com/1993/03/17/garden/can-a-green-candy-make-love-sweeter.html.

12. *General Python FAQ—Python 2.7.10 Documentation*, PYTHON.ORG, https://docs.python.org/2/faq/general.html (last visited June 20, 2015).

13. *Trademarks in Song Lyrics*, THE STRAIGHT DOPE (June 17, 2011), http://boards.straightdope.com/sdmb/showthread.php?t=612862.

14. For another Harry Potter case, see James Montgomery, *Canadian Folkies Can't Stop "Harry Potter," Judge Rules*, MTV NEWS (Nov. 8, 2005), http://www.mtv.com/news/articles/1513199/judge-canadian-group-cant-stop-harry-potter.jhtml.

Chapter 18: Trade Secrets

1. On trade secrets generally, see JERRY COHEN & ALAN S. GUTTERMAN, TRADE SECRETS: PROTECTION AND EXPLOITATION (1998).

2. DuPont de Nemours & Co. v. Kolon Indus. Inc., Nos. 10-1103, 10-1275 (4th Cir. 2011) (trade secrets in Kevlar fiber, with award to DuPont of $920 million).

3. For famous secret formulas, see Melanie Radzicki McManus, *10 Trade Secrets We Wish We Knew*, HOWSTUFFWORKS.COM, http://money.howstuffworks.com/10-trade-secrets1.htm (last visited June 21, 2015).

4. John Marsh, *Keep an Eye on Those Vendors: The Soft Underbelly of Trade Secret Protection*, THE TRADE SECRET LITIGATOR (June 24, 2011), http://www.tradesecretlitigator.com/2011/06/keep-an-eye-on-those-vendors-the-soft-underbelly-of-trade-secret-protection/.

5. Bruce Watson, *Shhh: 10 Make-or-Break Trade Secrets*, DAILYFINANCE (July 4, 2010, 3:00 PM), http://www.dailyfinance.com/2010/07/04/trade-secrets/.

6. Jenny Strasburg, *Lexar Wins Millions in Damages/Toshiba Ordered to Pay for Stealing Fremont Firm's Trade Secrets*, SFGATE (March 25, 2005), http://www.sfgate.com/business/article/Lexar-wins-millions-in-damages-Toshiba-ordered-2690176.php#ixzz252wPfqXg.

7. Learning Curve Toys, Inc. v. PlayWood Toys, 342 F.3d 714 (7th Cir. 2003).

8. Jacqui Cheng, *IBM's Siri Ban Highlights Companies' Privacy, Trade Secret Challenges*, ARS TECHNICA (May 23, 2010, 4:05 PM), http://arstechnica.com/apple/2012/05/ibms-siri-ban-highlights-companies-privacy-trade-secret-challenges/.

9. Gene Quinn, *Vault with Coca-Cola Trade Secret Formula on Public Display*, IPWATCHDOG (Jan. 6, 2010), http://www.ipwatchdog.com/2012/01/06/vault-with-coca-cola-trade-secret-formula-on-public-display.

10. McManus, *supra* note 3 (also quoting Cabrini College Assistant Professor of Business Scott Testa).

11. Seth Fiegerman, *The Biggest Apple Secrets Revealed during the First Week of the Patent Trial*, BUSINESS INSIDER (Aug. 6, 2012, 1:36 PM), http://www.businessinsider.com/10-apple-secrets-revealed-during-the-first-week-of-the-patent-trial-2012-8?op=1.

12. Goldin v. R.J. Reynolds Tobacco Co., 22 F. Supp. 61 (S.D.N.Y. 1938).

13. U.S. Patent Application No. PCT/US2004/037178 (filed Nov. 5, 2004).

Chapter 19: Trade Secrets and Employees

1. Brenda Goodman, *3 Arrested in Plot to Sell Coca-Cola Secrets to Rival*, INT'L HERALD TRIBUNE, July 6, 2006, http://www.nytimes.com/2006/07/06/business/worldbusiness/06iht-coke.2129199.html.

2. Jeanne C. Fromer, *Trade Secrecy in Willy Wonka's Chocolate Factory*, *in* THE LAW AND THEORY OF TRADE SECRECY: A HANDBOOK OF CONTEMPORARY RESEARCH 3 (Rochelle C. Dreyfuss & Katherine J. Strandburg eds., 2011).

3. ADRIAN JOHNS, PIRACY (2009).

4. Ronald J. Gilson, *The Legal Infrastructure of High Technology Industrial Districts: Silicon Valley, Route 128, and Covenants Not to Compete*, 74 N.Y. L. REV. 575 (1999). On the effects of restrictive employment agreements generally, see ORLY LOBEL, TALENT WANTS TO BE FREE: WHY WE SHOULD LEARN TO LOVE LEAKS, RAIDS, AND FREE RIDING (2013).

5. Evan Ackerman, *Battery Powered Battery Charger Is Probably Fake, Maybe Shouldn't Be,* OHGIZMO.COM, http://www.ohgizmo.com/2009/04/16/battery-powered-battery-charger-is-probably-fake-maybe-shouldnt-be/.

Chapter 20: International Intellectual Property

1. ADRIAN JOHNS, PIRACY 185 (2009) (quoting Robert Bell).

2. UNTAMED BEER, www.untamedbeer.com (last visited June 21, 2015).

3. GRAEME B. DINWOODIE, WILLIAM O. HENNESSEY, SHIRA PERLMUTTER & GRAEME W. AUSTIN, INTERNATIONAL INTELLECTUAL PROPERTY LAW AND POLICY 179, 237 (2d ed. 2008).

4. For stories on *Apple v. Samsung* litigation around the world, see *Apple v. Samsung/Ars Series*, ARS TECHNICA, http://arstechnica.com/series/apple-v-samsung/ (last visited June 21, 2015).

5. DINWOODIE ET AL., *supra* note 3, at 519.

6. PAT CHOATE, HOT PROPERTY: THE STEALING OF IDEAS IN AN AGE OF GLOBALIZATION (2005).

7. Teresa Riordan, *Patents: An Economist Strolls through History and Turns Patent Theory Upside Down*, N.Y. TIMES, Sept. 29, 2003, http://www.nytimes.com/2003/09/29/business/patents-an-economist-strolls-through-history-and-turns-patent-theory-upside-down.html (quoting Petra Moser).

8. *The Wright Brothers & the Invention of the Aerial Age* (exhibition description), http://nasm.si.edu/wrightbrothers/ (last visited June 21, 2015).

9. *See* John F. Duffy, *Rethinking the Prospect Theory of Patents*, 71 U. CHI. L. REV. 439 (2004).

10. The mission statement for the Smithsonian Institution's National Air and Space Museum begins with "the original 1903 Wright Flyer" and ends with "the sole-surviving Boeing 307 Stratoliner and space shuttle *Enterprise.*" *National Air and Space Museum Press Kit*, http://airandspace.si.edu/about/newsroom/press-kits.cfm (last visited Oct. 14, 2010).

11. *See* http://www.paperclipcampaign.com/why-a-paperclip. Norway's paperclip was not the most popular—and its role as a national symbol is a little unclear (Norway's heroic resistance is well documented). For more information on Norwegian invention lore, see NORTH, http://www.blather.net/north/archives/norwegian_inventions/ (last visited June 21, 2015) (along with the treadle pump, the cheese slicer, and a method of opening a beer bottle with a lighter).

12. David Adam, *Can Nobody Make a Sandwich Like McDonald's?*, THE GUARDIAN, Nov. 19, 2006, http://www.theguardian.com/lifeandstyle/2006/nov/20/foodanddrink.uk.

13. Cumberland Pencil Museum, Keswick, Cumbria.

14. THE THIRD MAN (London Film Productions 1949) (character Harry Lime speaking).

15. *The Joke Show*, A PRAIRIE HOME COMPANION (April 17, 2004).

16. Diamond v. Chakrabarty, 447 U.S. 303, 316 n.10 (1980) (listing "telegraph (Morse, No. 1,647); telephone (Bell, No. 174,465); electric lamp (Edison, No. 223,898); airplane (the Wrights, No. 821,393); transistor (Bardeen & Brattain, No. 2,524,035); neutronic reactor (Fermi & Szilard, No. 2,708,656); laser (Schawlow & Townes, No. 2,929,922)").

17. *See* Marconi Wireless Tel. Co. v. United States, 320 U.S. 1, 4 (1943).

18. JOHNS, *supra* note 1, at 100 (discussing 18th-century druggists).

19. *Id.* at 174.

20. THE PATRY COPYRIGHT BLOG, http://williampatry.blogspot.com/ (last visited June 21, 2015).

21. *Robin Williams: 50 Great Quotes*, THE TELEGRAPH, http://www.telegraph.co.uk/culture/culturepicturegalleries/11027564/Robin-Williams-50-great-quotes.html?frame=3003238 (last visited July 26, 2015) ("In England, if you commit a crime, the police don't have a gun and you don't have a gun. If you commit a crime, the police will say, 'Stop, or I'll say stop again'").

22. U.S. COPYRIGHT OFFICE, INTERNATIONAL COPYRIGHT RELATIONS OF THE UNITED STATES (2010), *available at* http://www.copyright.gov/circs/circ38a.pdf.

23. JULIE E. COHEN & LYDIA P. LOREN, COPYRIGHT IN A GLOBAL INFORMATION ECONOMY (3rd ed. 2010).

24. Christoph Engel & Michael Kurschilgen, Fairness Ex Ante and Ex Post: An Experimental Test of the German "Bestseller Paragraph," MPI Collective Goods Preprint, No. 2010/29.

25. THOMAS COTTER, COMPARATIVE PATENT REMEDIES: A LEGAL AND ECONOMIC ANALYSIS (2013).

26. These four paragraphs rely greatly on B. ZORINA KHAN, THE DEMOCRATIZATION OF INVENTION: PATENTS AND COPYRIGHTS IN AMERICAN ECONOMIC DEVELOPMENT, 1790–1920, at 53–64, 124–29 (2005).

27. *Taking Pains: Indian Patent Rules Infuriate Big Pharma*, THE ECONOMIST, Sept. 8, 2012, http://www.economist.com/node/21562226.

28. Kyle Orland, *Microsoft Patent Application Shows Holodeck-Style "Immersive Display,"* ARS TECHNICA (Sept. 12, 2012), http://arstechnica.com/gaming/2012/09/microsoft-patent-shows-holodeck-style-full-room-immersive-display/.

29. Jeffery Cyril Atik & Hans Henrik Lidgard, *Embracing Price Discrimination—TRIPS and Parallel Trade in Pharmaceuticals*, 28 U. PA. J. INT'L ECON. L. 1043 (2006).

For further reading:

For information on industry standards, see Christopher Gibson, *Globalization and the Technology Standards Game: Balancing Concerns of Protectionism and Intellectual Property in International Standards*, 22 BERKELEY TECH. L.J. 1401 (2007).

For a forward-looking perspective, see Mary Wai San Wong, The Next Ten Years in Copyright Law: An Asian Perspective (Feb. 2007), *available at* http://dx.doi.org/10.2139/ssrn.1017144.

Culture skirmishes are always interesting. Disputes over ownership likewise may involve the original works, such as the long-running dispute between Britain and Greece over the statues, now in the British Museum, known as the "Elgin Marbles" or the "Parthenon Marbles." *See generally* IMPERIALISM, ART AND RESTITUTION (John Henry Merryman ed., 2006).

As a good example of the slightly nationalistic tendencies of museums discussed in the chapter, the computer exhibit in the Science Museum in London contains not just material on computing pioneers like Alan Turing, but also on punch cards of the British Tabulating Machine Company Limited, which get less emphasis elsewhere.

Chapter 21: Indigenous Peoples and Intellectual Property

1. This chapter draws from Lorie Graham & Stephen McJohn, *Indigenous Peoples and Intellectual Property*, 19 WASH. U. J.L. & POL'Y 313 (2006).

2. *Followup to: The Phrase "Wampum Jewelry,"* THEBEADSITE.COM (July 18, 2000), http://www.thebeadsite.com/wampname.htm.

3. Matthew L. M. Fletcher, *Navajo Nation Sues Urban Outfitters for Trademark Violations*, TURTLE TALK (Feb. 29, 2012), http://turtletalk.wordpress.com/2012/02/29/navajo-nation-sues-urban-outfitters-for-trademark-violations/.

4. John Vidal, *Biopirates Who Seek the Greatest Prizes*, THE GUARDIAN, Nov. 14, 2000, http://www.guardian.co.uk/science/2000/nov/15/genetics2 (complete with British punctuation).

5. Deepa Varadarajan, *A Trade Secret Approach to Protecting Traditional Knowledge*, 36 YALE J. INT'L L. 372 (2011); Shayana Kadidal, *Plants, Poverty, and Pharmaceutical Patents*, 103 YALE L.J. 223 (1993).

6. Peter K. Yu, *Cultural Relics, Intellectual Property, and Intangible Heritage*, 81 TEMP. L. REV. 433 (2008).

7. *See* Varadarajan, *supra* note 5.

8. DARRELL A. POSEY & GRAHAM DUTFIELD, BEYOND INTELLECTUAL PROPERTY: TOWARD TRADITIONAL RESOURCE RIGHTS FOR INDIGENOUS PEOPLES AND LOCAL COMMUNITIES (1996).

9. *See* Varadarajan, *supra* note 5.

10. Suketu Mehta, *Can You Patent Wisdom?*, N.Y. TIMES, May 7, 2007, at http://www.nytimes.com/2007/05/07/opinion/07iht-edmehta.1.5596253.html?_r=0.

11. *Heilbrunn Timeline of Art History: African Influences on Modern Art*, THE METROPOLITAN MUSEUM OF ART, http://www.metmuseum.org/toah/hd/aima/hd_aima.htm (last visited June 22, 2015) ("In France, Henri Matisse, Pablo Picasso, and their School of Paris friends blended the highly stylized treatment of the human figure in African sculptures with painting styles derived from the post-Impressionist works of Cézanne and Gauguin.").
    The Metropolitan Museum in New York provides any number of examples of indigenous art that prefigures 20th-century art of several schools (in addition to showing how artists borrowed from their contemporaries and precursors in the same genre, as discussed earlier in the book).

12. *Id.*

13. Angela R. Riley, *Recovering Collectivity*, 18 CARDOZO ARTS & ENT. L.J. 175, 177 (2000).

14. Lorie Graham, *Securing Economic Sovereignty through Agreement*, 37 NEW ENG. L. REV. 523, 537 (2003) (discussing factors a tribe might consider in adopting commercial laws and adapting them to the tribe's own norms); Charles R. McManis, *Intellectual Property, Genetic Resources and Traditional Knowledge Protection: Thinking Globally, Acting Locally*, 11 CARDOZO J. INT'L & COMP. L. 547, 553 (2003) (discussing how international initiatives toward protection of traditional knowledge and folklore have contributed to the making of local agreements "among research institutions, companies, communities and non-governmental organizations").

15. Erica-Irene Daes, *Protection of the Heritage of Indigenous People*, U.N. Subcomm'n on Prevention of Discrimination and Prot. of Minorities, U.N. Doc. E.97.XIV.3, PP 21-32 (1997); *see also* Resolution of Subcomm'n on Prevention of Discrimination and Prot. of Minorities, E/CN.4/Sub.2/1992/30, 1992/35 (Aug. 27, 1992).

16. Declaration on the Rights of Indigenous Peoples, G.A. Res. 61/295, U.N. Doc. A/RES/61/295, Article 31 (Sept. 13, 2007).

17. MICHAEL F. BROWN, WHO OWNS NATIVE CULTURE? (2003) (lucidly raises concerns about overprotection); www.williams.edu/go/nativesupplements provides access to a number of related research resources on overprotection of traditional knowledge.

18. For resources on reasons needed to protect traditional knowledge, cultural property, folklore, and other indigenous knowledge, see S. JAMES ANAYA, INTERNATIONAL HUMAN RIGHTS AND INDIGENOUS PEOPLES (2009); JAMES BOYLE, SHAMANS, SOFTWARE, AND SPLEENS (1996); Keith Aoki, *Neocolonialism, Anticommons Property, and Biopiracy in the (Not-So-Brave) New World Order of International Intellectual Property Protection*, 6 IND. J. GLOBAL LEGAL STUD. 11 (1998); RUSSEL L. BARSH, FIRST PEOPLES WORLDWIDE, THE NORTH AMERICAN PHARMACEUTICAL INDUSTRY AND RESEARCH INVOLVING INDIGENOUS KNOWLEDGE (2001); Shubha Ghosh, *Reflections on the Traditional Knowledge Debate*, 11 CARDOZO J. INT'L & COMP. L. 497, 497 (2003); Daniel J. Gervais, *Spiritual but Not Intellectual? The Protection of Sacred Intangible Traditional Knowledge*, 11 CARDOZO J. INT'L & COMP. L. 467, 494–95 (2003); Megan M. Carpenter, *Intellectual Property Law and Indigenous Peoples: Adapting Copyright Law to the Needs of a Global Community*, 7 YALE HUM. RTS. & DEV. L.J. 51–78 (2004); Angela R. Riley, *"Straight Stealing": Towards an Indigenous System of Cultural Property Protection*, 80 WASH. L. REV. 69 (2005).

19. Rosemary J. Coombe, *The Recognition of Indigenous Peoples' and Community Traditional Knowledge in International Law*, 14 ST. THOMAS L. REV. 275 (2001).

20. *See* Graham & McJohn, *supra* note 1.

21. Tribal knowledge is not protected, unlike trade secrets. *See* BUREAU OF LAND MGMT., GENERAL PROCEDURAL GUIDANCE FOR NATIVE AMERICAN CONSULTATION, REP. NO. H-8160-1, Ch. III, Sec. F, *available at* http://www.blm.gov/style/medialib/blm/wo/Information_Resources_Management/policy/blm_handbook.Par.38741.File.dat/H-8120-1.pdf ("One of the greatest barriers to completely open consultation discussions is Native Americans'

hesitation to divulge information about places that are considered to have a sacred character, or practices that are of a sacred or private nature. . . . We must not overstate our ability to protect sensitive information.").

22. Stephen M. McJohn, *Eldred's Aftermath*, 10 Mich. Telecomm. & Tech. L. Rev. 95 (2003); Pamela Samuelson, *The Copyright Grab*, 4.01 Wired, Jan. 1996, at 134; Mark A. Lemley, *The Modern Lanham Act and the Death of Common Sense*, 108 Yale L.J. 1687 (1999); Kenneth L. Port, *The Congressional Expansion of American Trademark Law*, 35 Wake Forest L. Rev. 827 (2000).

23. Anupam Chander & Madhavi Sunder, *The Romance of the Public Domain*, 92 Cal. L. Rev. 1331 (2004).

24. *See* Graham, *supra* note 14.

25. Paul Kuruk, *Protecting Folklore under Modern Intellectual Property Regimes: A Reappraisal of the Tensions between Individual and Communal Rights in Africa and the United States*, 48 Am. U.L. Rev. 769, 819–22 (1999).

26. *See* Lorie Graham, *A Right to Media?*, 41 Colum. Human Rights L. Rev. 429 (2010); Wallace Coffey & Rebecca Tsosie, *Rethinking the Tribal Sovereignty Doctrine: Cultural Sovereignty and the Collective Future of Indian Nations*, 12 Stan. L. & Pol'y Rev. 191 (2001).

27. Chander & Sunder, *supra* note 23.

28. See Kristen A. Carpenter, Sonia K. Katyal & Angela R. Riley, *In Defense of Property*, 118 Yale L.J. 1022, 1103 (2009) ("[A]ffording indigenous groups even minimum protections and profit-sharing rights in harvesting, collecting, organizing, disseminating, and selling their traditional knowledge is crucial, and it can be achieved without employing the absolute ownership rights or exclusive access that cultural property critics fear.").

29. Quileute Oceanside Resort, http://www.quileuteoceanside.com/ (last visited June 22, 2015); Angela R. Riley, *Sucking the Quileute Dry*, N.Y. Times, Feb. 7, 2010, http://www.nytimes.com/2010/02/08/opinion/08riley.html.

30. Glacier Nat'l Park's Sun Tours, http://www.glaciersuntours.com/ (last visited June 22, 2015).

31. Swaraj Paul Barooah, *Traditional Knowledge Patent Applications: Need for Deliberation*, SpicyIP (Dec. 23, 2012), http://spicyip.com/2012/12/guest-post-traditional-knowledge-patent.html (quoting WIPO, Intellectual Property and Traditional Knowledge, Booklet No. 2).

32. Graham Dutfield, *TRIPS-Related Aspects of Traditional Knowledge*, 33 Case W. Res. J. Int'l L. 233, 273 (2001) ("It seems highly unlikely that a new framework to protect TK will be inserted into TRIPS anytime soon. And since the United States is determined to prevent a WIPO convention on TK that could then be incorporated in TRIPS, this is unlikely to happen even in the more distant future. At best, minimalist measures to safeguard TK from misappropriation could conceivably be agreed upon. A greater danger is that trade negotiators will sacrifice the interests of traditional knowledge holders once concessions in other areas of intellectual property or other trade-related issues are secured in return.").

For further reading:

For reading on the Native American Graves Protection and Repatriation Act (NAGPRA), see Suzan Shown Harjo, *Introduction* to American Indian Ritual Object Repatriation Foundation, Mending the Circle: A Native American Repatriation Guide 3 (1996); Angela R. Riley, *Indian Remains, Human Rights*, 34 Colum. Hum. Rts. L. Rev. 49 (2002); Kristen A. Carpenter, *Property Rights Approach to Sacred Sites Cases: Asserting a Place for Indians as Non-Owners*, 52 UCLA L. Rev. 1061 (2005).

The World Intellectual Property Organization (WIPO) is always looking at indigenous issues. An Intergovernmental Committee on Intellectual Property and Genetic Resources, Traditional Knowledge and Folklore, has been established in connection with WIPO. *See* http://www.wipo.int/ (last visited July 26, 2015).

On the issues that may arise in negotiations to protect indigenous people's information, see Caroline Joan S. Picart, *Cross Cultural Negotiations and International Intellectual Property Law: Attempts to Work across Cultural Clashes between Indigenous Peoples and Majoritarian Cultures*, 23 S. Cal. Interdisc. L.J. 37–65 (2014) (quoting David Livermore, Leading with Cultural Intelligence: The New Secret to Success 106 (2010)).

An extensive collection of indigenous art, along with notes on its influence on others, is at the Musée du Quay Branly, Paris.

1. This chapter is based on JAY DRATLER & STEPHEN M. MCJOHN, CYBERLAW: INTELLECTUAL PROPERTY IN THE DIGITAL MILLENNIUM (2015).

2. Mark Frauenfelder, *The Secret History of the St. Louis Post Office and Its Amazing Pneumatic Tube*, BOING BOING (Sept. 5, 2012), http://boingboing.net/2012/09/05/the-secret-history-of-the-st.html (discussing Aimee Levitt's article in *Riverfront Times*).

3. BENJAMIN KAPLAN, AN UNHURRIED VIEW OF COPYRIGHT 119 (1967).

4. *Id.* at 120.

5. *Id.*

6. *Id.* at 122.

7. *Id.* at 123.

8. *Takedown Hall of Shame*, ELEC. FRONTIER FOUND., https://www.eff.org/takedowns (last visited June 22, 2015).

9. Timothy B. Lee, *YouTube Finally Offers a Meaningful ContentID Appeal Process*, ARS TECHNICA (Oct. 3, 2012), http://arstechnica.com/tech-policy/2012/10/youtube-finally-offers-a-meaningful -contentid-appeal-process/.

10. Cory Doctorow, *Curiosity Landing Removed from YouTube after Bogus Copyright Claim by Scripps*, BOING BOING (Aug. 6, 2012), http://boingboing.net/2012/08/06/curiosity-landing-removed-from.html.

11. THE NEW HACKER'S DICTIONARY (Eric S. Raymond ed., 2d ed. 1993).

12. PINGDOM ROYAL, www.royal.pingdom.com (last visited June 22, 2015).

13. *Id.*

14. Antonio Regalado, *Nathan Myhrvold's Cunning Plan to Prevent 3-D Printer Piracy*, MIT TECH. REV. (Oct. 11, 2012), http://www.technologyreview.com/view/429566/nathan-myhrvolds-cunning-plan-to-prevent-3-d -printer-piracy/.

15. Michael Rustad & Diane D'Angelo, *The Path of Internet Law: An Annotated Guide to Legal Landmarks*, 2011 DUKE L. & TECH. REV. 12 (2011).

16. Loek Essers, *Google Denied Claim to Oogle.com Domain Name*, PCWORLD (July 26, 2012), http://www .pcworld.com/article/259891/google_denied_claim_to_ooglecom_domain_name.html.

For further reading:

Regulatory agencies may affect how open the Internet remains to innovation. *See* SUSAN CRAWFORD, CAPTIVE AUDIENCE: THE TELECOM INDUSTRY AND MONOPOLY POWER IN THE NEW GILDED AGE (2013).

For more on hazards to innovation, from DRM and other controls, see JONATHAN ZITTRAIN, THE FUTURE OF THE INTERNET—AND HOW TO STOP IT (2009).

For a clear discussion of the background of domain name arbitration and key issues, see Christopher Gibson, *Uniform Domain Name Dispute Resolution Policy (UDRP)*, WORLD INTELLECTUAL PROPERTY ORGANIZATION.

Index

About the ABA Section of Intellectual Property Law

From its strength within the American Bar Association, the ABA Section of Intellectual Property Law (ABA-IPL) advances the development and improvement of intellectual property laws and their fair and just administration. The Section furthers the goals of its members by sharing knowledge and balanced insight on the full spectrum of intellectual property law and practice, including patents, trademarks, copyright, industrial design, literary and artistic works, scientific works, and innovation. Providing a forum for rich perspectives and reasoned commentary, ABA-IPL serves as the ABA voice of intellectual property law within the profession, before policy makers, and with the public.

ABA Section of Intellectual Property Law
Order today! Call 1-800-285-2221
Monday-Friday, 7:30 a.m. – 5:30 p.m., Central Time
or Visit the ABA Web Store: www.ShopABA.org

Qty	Title	Regular Price	ABA-IPL Member Price	Total
_____	ADR Advocacy, Strategies, and Practice in Intellectual Property Cases (5370195)	$139.95	$114.95	$_____
_____	ANDA Litigation (5370199)	$299.00	$249.00	$_____
_____	Careers in IP Law (5370204)	$24.95	$16.95	$_____
_____	Computer Games and Virtual Worlds (5370172)	$69.95	$55.95	$_____
_____	Copyright Remedies (5370208)	$89.95	$74.95	$_____
_____	Distance Learning and Copyright (5370163)	$89.95	$79.95	$_____
_____	Fundamentals of Intellectual Property Valuation (5370143)	$69.95	$49.95	$_____
_____	IP Attorney's Handbook for Insurance Coverage in Intellectual Property Disputes, Second Edition (5370210)	$139.95	$129.95	$_____
_____	IP Protection in China (5370217)	$139.95	$109.95	$_____
_____	A Lawyer's Guide to Section 337 Investigations before the U.S. International Trade Commission, Second Edition (5370203)	$119.95	$89.95	$_____
_____	A Legal Strategist's Guide to Trademark Trial and Appeal Board Practice, Second Edition (5370200)	$159.95	$129.95	$_____
_____	Music & Copyright in America (5370201)	$97.95	$67.95	$_____
_____	New Practitioner's Guide to Intellectual Property (5370198)	$89.95	$69.95	$_____
_____	The Patent Infringement Litigation Handbook (1620416)	$149.95	$129.95	$_____
_____	Patently Persuasive (5370206)	$129.95	$99.95	$_____
_____	Patent Obviousness in the Wake of KSR International Co. v. Teleflex Inc. (5370189)	$129.95	$103.95	$_____
_____	Patent Trial Advocacy Casebook, Third Edition (5370124)	$149.95	$119.95	$_____
_____	Practitioner's Guide to the PCT (5370194)	$139.95	$109.95	$_____
_____	Practitioner's Guide to Trials Before the Patent Trial and Appeal Board (5370209)	$139.95	$114.95	$_____
_____	Pre-ANDA Litigation (5370212)	$275.00	$220.00	$_____
_____	Preliminary Relief in Patent Infringement Disputes (5370194)	$119.95	$94.95	$_____
_____	Right of Publicity (5370215)	$89.95	$74.95	$_____
_____	Settlement of Patent Litigation and Disputes (5370192)	$179.95	$144.95	$_____
_____	Starting an IP Law Practice (5370202)	$54.95	$34.95	$_____
_____	The Tech Contracts Handbook, Second Edition (5370216)	$39.95	$34.95	$_____
_____	Technology Transfer Law Handbook (5370211)	$220.00	$176.00	$_____
_____	Trademark and Deceptive Advertising Surveys (5370197)	$179.95	$134.95	$_____
_____	Trademark Surveys (5370207)	$269.95	$239.95	$_____

* Tax	**Payment**
DC residents add 5.75%	❏ Check enclosed payable to the ABA
IL residents add 9.25%	❏ VISA ❏ Mastercard ❏ American Express

* Tax	$_____
** Shipping/Handling	$_____
TOTAL	$_____

**Shipping/Handling
Up to $49.99.................. $5.95
$50 to $99.99 .... $7.95
$100 to $199.99.............. $9.95
$200 to $499.99............. $12.95
$500 to $999.99............. $15.95
$1,000 and above........... $18.95

Name_____

Firm/Organization_____

Address_____

City_____ State_____ Zipcode_____

Phone_____ E-mail_____
(in case of questions about your order)

Please allow 5 to 7 business days for UPS delivery. Need it sooner? Ask about overnight delivery. Call the ABA Service Center at 1-800-285-2221 for more information.

Guarantee: If – for any reason – you are not satisfied with your purchase, you may return it within 30 days of receipt for a complete refund of the price of the book(s). No questions asked!

Please mail your order to:
ABA Publication Orders, P.O. Box 10892, Chicago, Illinois 60610-0892
Phone: 1-800-285-2221 or 312-988-5522 • Fax: 312-988-5568
E-mail: orders@abanet.org

Thank you for your order!

Rights of Publicity

Exclusive right to exclude others from commercializing or $profit by use of another person's being

Want to protect against:
- false endorsments // Right to choose what they endorse
- over exposure // Right to control media exposure (to an extent) & privacy

State - Based Laws
- Not Every state has these laws
- Some Fed laws as well

If sueing :
1. Commercial Use
2. Without Consent   } Plaintiff need to prove
3. For Trade Purposes

Duration of Right varries State to State.
- ° At death
- ° death + 10 yrs
- ° death + 100 yrs (Oklahoma)
- ° death + 70 yrs (California)

} Where the use is occuring } Multiple States can become an option if the use is national

Case: ~~Heres~~ Carson v. Here's Johnny

Using Carson's name in reference to a well known phrase on Carson's show

Carson had given other businesses the right to use the phrase, Liscensed to them

Here's Johnny (D) did not seek permission from Carson

Court: Misapropriation of Carson's identity

Lanham Act:

Federal Trademark Act in 1946

Prohibits:
- Trademark Infringement
- Trademark dilution
- False Advertising